REPUBLIC OF SOUTH AFRICA
REPUBLIEK VAN SUID-AFRIKA

DEPARTMENT OF AGRICULTURE
DEPARTEMENT VAN LANDBOU

FLORA OF SOUTHERN AFRICA

VOLUME 7

PART 2,   FASCICLE 2

ISBN 0 621 07939 1

G.P.-S

FLORA OF SOUTHERN AFRICA

which deals with the territories of

SOUTH AFRICA, CISKEI, TRANSKEI, LESOTHO, SWAZILAND, BOPHUTHA-
TSWANA, SOUTH WEST AFRICA/NAMIBIA, BOTSWANA AND VENDA

VOLUME 7   IRIDACEAE

PART 2   IXIOIDEAE

Fascicle 2 Syringodea, Romulea

by

Miriam P. de Vos

Edited by

O.A. Leistner

Editorial Committee: B. de Winter, D. J. B. Killick and O. A. Leistner

Botanical Research Institute,
Department of Agriculture

1983

CONTENTS

NEW TAXA, NEW COMBINATION AND NEW NAME PUBLISHED IN FASCICLE 2 PART 2

Romulea luteoflora *(De Vos) De Vos* var. **sanisensis** *De Vos,* var. nov., p. 2.2: 56
Romulea vlokii *De Vos,* sp. nov., p. 2,2: 55
Romulea neglecta *(Schultes) De Vos,* comb. nov., p. 2,2: 29
Syringodea bifucata *De Vos,* nom. nov., p. 2,2: 4

Date of Publication: April 1983

INTRODUCTION

For a key to the families and the genera not keyed out in this part, the Flora should be used in conjunction with Dyer's Genera of Southern African Flowering Plants, Vol. 1 (1975) and Vol. 2 (1976), which are arranged on the lines of the Engler system. The genera are numbered, as far as possible, according to the list published by De Dalla Torre and Harms in their Genera Siphonogamarum (1900-1907) in order to facilitate reference, though genera in the Flora are not necessarily arranged in this sequence.

The following condensed abbreviations for literature references are used:

C.F.A.	Conspectus Florae Angolensis
R. A. Dyer, Gen.	The Genera of Southern African Flowering Plants by R. A. Dyer, Vol. 1 (1975) and Vol. 2 (1976)
F.C.	Flora Capensis
F.C.B.	Flore du Congo et du Rwanda-Burundi
F.S.W.A.	Prodromus einer Flora von Südwestafrika
F.T.A.	Flora of Tropical Africa
F.T.E.A.	Flora of Tropical East Africa
F.W.T.A.	Flora of West Tropical Africa
F.Z.	Flora Zambesiaca
Burtt Davy, Fl. Transv. ..	Manual of the Flowering Plants and Ferns of the Transvaal and Swaziland, Vol. 1 (1926) and Vol. 2 (1932).

Cited voucher specimens given without indication of herbarium are housed in PRE (National Herbarium, Pretoria).

Localities are sometimes referred to in terms of the degree reference system (Leistner & Morris in Ann. Cape Prov. Mus. 12: 1 − 565; 1976).

This part was compiled in accordance with a Guide to Contributors to the Flora of Southern Africa (Ross, Leistner & De Winter, 1977), which is available from the Librarian, Botanical Research Institute, Private Bag X101, Pretoria, 0001.

Volume 7 of the Flora, of which the present publication is a component, will appear in two parts of which the second is divided into two fascicles (see p. viii). The number of the part, which in the present publication is '2' and the number of the fascicle, again '2', precede the page number on all pages marked with Arabic numerals. This was done with a view to binding the entire volume, once completed, and to compiling a combined index to all its component parts. When binding the entire volume the pages marked with Roman numerals may be omitted.

PLAN OF FLORA OF SOUTHERN AFRICA

Cryptogam volumes will in future not be numbered but will be known by the name of the group they cover. The number assigned to the volume on Characeae therefore becomes redundant.

Alien families are marked with an asterisk.

Published volumes and parts are shown in italics.

Please note that local prices as given below do not include GST.

INTRODUCTORY VOLUMES

The genera of Southern African flowering plants
    Vol. 1: *Dicotyledons* (Published 1975). Price: R11,00. Overseas: R14,00. Post free
    Vol. 2: *Monocotyledons* (Published 1976). Price: R8,00. Overseas: R10,00. Post free

Botanical exploration in Southern Africa

CRYPTOGAM VOLUMES

Charophyta (Published as Vol. 9 in 1978). Price: R4,25. Overseas: R5,30. Post free

Bryophyta:

        Part 1: Mosses: Fascicle 1: *Sphagnaceae — Grimmiaceae* (Published 1981). Price: R24,33, GST R0,97
                        Total R25,30. Overseas: R30,40. Post free
                Fascicle 2: Gigaspermaceae — Bartramiaceae
                Fascicle 3: Erpodiaceae — Hookeriaceae
                Fascicle 4: Fabroniaceae — Polytrichaceae
        Part 2: Hepatophyta, Anthocerotophyta

Pteridophyta

FLOWERING PLANTS VOLUMES

Vol. 1: *Stangeriaceae, Zamiaceae, Podocarpaceae, Pinaceae*, Cupressaceae, Welwitschiaceae, Typhaceae, Zosteraceae, Potamogetonaceae, Ruppiaceae, Zannichelliaceae, Najadaceae, Aponogetonaceae, Juncaginaceae, Alismataceae, Hydrocharitaceaᵉ* (Published 1966). Price: R1,75. Overseas: R2,20. Post free

Vol. 2: Poaceae

Vol. 3: Cyperaceae, Arecaceae, Araceae, Lemnaceae, Flagellariaceae

Vol. 4: Restionaceae, Mayacaceae, Xyridaceae, Eriocaulaceae, Commelinaceae, Pontederiaceae, Juncaceae

Vol. 5: Liliaceae, Agavaceae

Vol. 6: Haemodoraceae, Amaryllidaceae, Hypoxidaceae, Tecophilaeaceae, Velloziaceae, Dioscoreaceae

Vol. 7: Iridaceae: Part 1: Nivenioideae, Iridoideae
                Part 2: Ixioideae: Fascicle 1
                                Fascicle 2: *Syringodea, Romulea* (Published 1983).
                                        Price: R4,20. Overseas: R5,00.
                                        Post free

Vol. 8: Musaceae, Strelitziaceae, Zingiberaceae, Cannaceae*, Burmanniaceae, Orchidaceae

Vol. 9: Casuarinaceae*, Piperaceae, Salicaceae, Myricaceae, Fagaceae*, Ulmaceae, Moraceae, Cannabaceae*, Urticaceae, Proteaceae

Vol. 10: Part 1: *Loranthaceae, Viscaceae* (Published 1979). Price: R2,00. Overseas: R2,50. Post free
Santalaceae, Grubbiaceae, Opiliaceae, Olacaceae, Balanophoraceae, Aristolochiaceae, Rafflesiaceae, Hydnoraceae, Polygonaceae, Chenopodiaceae, Amaranthaceae, Nyctaginaceae

Vol. 11: Phytolaccaceae, Aizoaceae, Mesembryanthemaceae

Vol. 12: Portulacaceae, Basellaceae, Caryophyllaceae, Illecebraceae, Cabombaceae, Nymphaeaceae, Ceratophyllaceae, Ranunculaceae, Menispermaceae, Annonaceae, Trimeniaceae, Lauraceae, Hernandiaceae, Papaveraceae, Fumariaceae

Vol. 13: *Brassicaceae, Capparaceae, Resedaceae, Moringaceae, Droseraceae, Roridulaceae, Podostemaceae, Hydrostachyaceae* (Published 1970). Price: R10,00. Overseas: R12,00. Post free

Vol. 14: Crassulaceae

Vol. 15: Vahliaceae, Montiniaceae, Escalloniaceae, Pittosporaceae, Cunoniaceae, Myrothamnaceae, Bruniaceae, Hamamelidaceae, Rosaceae, Connaraceae

Vol. 16: Fabaceae: Part 1: *Mimosoideae* (Published 1975). Price: R13,50. Overseas: R16,75. Post free
Part 2: *Caesalpinioideae* (Published 1977). Price: R16,00. Overseas: R20,00. Post free

       Papilionoideae

Vol. 17: Geraniaceae, Oxalidaceae

Vol. 18: Linaceae, Erythroxylaceae, Zygophyllaceae, Balanitaceae, Rutaceae, Simaroubaceae, Burseraceae, Ptaeroxylaceae, Meliaceae, Aitoniaceae, Malpighiaceae

Vol. 19: Polygalaceae, Dichapetalaceae, Euphorbiaceae, Callitrichaceae, Buxaceae, Anacardiaceae, Aquifoliaceae

Vol. 20: Celastraceae, Icacinaceae, Sapindaceae, Melianthaceae, Greyiaceae, Balsaminaceae, Rhamnaceae, Vitaceae

Vol. 21: Tiliaceae, Malvaceae, Bombacaceae, Sterculiaceae

Vol. 22: *Ochnaceae, Clusiaceae, Elatinaceae, Frankeniaceae, Tamaricaceae, Canellaceae, Violaceae, Flacourtiaceae, Turneraceae, Passifloraceae, Achariaceae, Loasaceae, Begoniaceae, Cactaceae* (Published 1976). Price: R8,60. Overseas: R10,75. Post free

Vol. 23: Geissolomaceae, Penaeaceae, Oliniaceae, Thymelaeaceae, Lythraceae, Lecythidaceae

Vol. 24: Rhizophoraceae, Combretaceae, Myrtaceae, Melastomataceae, Onagraceae, Trapaceae, Haloragaceae, Gunneraceae, Araliaceae, Apiaceae, Cornaceae

Vol. 25: Ericaceae

Vol. 26: *Myrsinaceae, Primulaceae, Plumbaginaceae, Sapotaceae, Ebenaceae, Oleaceae, Salvadoraceae, Loganiaceae, Gentianaceae, Apocynaceae* (Published 1963). Price: R4,60. Overseas: R5,75. Post free

Vol. 27: Part 1: Periplocaceae, Asclepiadaceae (Microloma-Xysmalobium)
Part 2: Asclepiadaceae (Schizoglossum-Woodia)
Part 3: Asclepiadaceae (Asclepias-Anisotoma)
Part 4: *Asclepiadaceae (Brachystelma-Riocreuxia)* (Published 1980). Price: R4,50. Overseas: R6,00. Post free

       Asclepiadaceae (remaining genera)

Vol. 28: Cuscutaceae, Convolvulaceae, Hydrophyllaceae, Boraginaceae, Stilbaceae, Verbenaceae, Lamiaceae, Solanaceae, Retziaceae

Vol. 29: Scrophulariaceae

Vol. 30: Bignoniaceae, Pedaliaceae, Martyniaceae, Orobanchaceae, Gesneriaceae, Lentibulariaceae, Acanthaceae, Myoporaceae

Vol. 31: Plantaginaceae, Rubiaceae, Valerianaceae, Dipsacaceae, Cucurbitaceae

Vol. 32: Campanulaceae, Sphenocleaceae, Lobeliaceae, Goodeniaceae

Vol. 33: Asteraceae

1260   **SYRINGODEA**

Syringodea *Hook. f.* in Curtis's bot. Mag. 29: t. 6072 (1873), nom. conserv. (non *Syringodea* D. Don., 1834); Bak. in J. Bot., Lond. 5: 66 (1876); in J. Linn. Soc., Bot. 16: 85 (1877); Handb. Irid. 95 (1892); in F.C. 6: 34 (1896); Benth. & Hook. f., Gen. Pl. 3: 693 (1883); Klatt in Abh. Naturforsch. Ges. Halle 15: 403 (1882); in Dur. & Schinz, Consp. Fl. Afr. 5: 160 (1895); Pax in Natürl. PflFam. 2, 5: 475 (1888); Kuntze, Rev. Gen. 3: 309 (1898); Diels in Natürl. PflFam. edn 2, 15a: 463 (1930); Phill., Gen. 212 (1951); Goldbl. in Jl S. Afr. Bot. 37: 393 (1971); in R. A. Dyer, Gen. 2: 965 (1976); De Vos in Jl S. Afr. Bot. 40: 201 (1974). Type species: *S. pulchella* Hook. f.

Plants small, deciduous. *Corm* small, asymmetrical, almost turbinate or obovoid or laterally flattened, usually with an obliquely ridged base; tunics entire, woody or membranous, split into fine parallel fibrils on the basal ridge and at the top into long pointed teeth. *Stem* very short, subterranean, hidden by leaf bases. *Cataphylls* usually 2, membranous, sheathing the base of the shoot, the tips appearing above-ground. *Foliage leaves* several or rarely one, basal, largely bifacial, filiform with an adaxial groove, or lanceolate or linear. *Inflorescence* with one or more flowers appearing successively, each terminal on a very short subterranean peduncle and enclosed at its base by a 2-valved spathe. *Spathe valves (bracts)* largely membranous, subhyaline, greenish in the upper half. *Flowers* actinomorphic, funnel-shaped to salver-shaped, thermonastic. *Perianth tube* narrowly tubular, erect, longer than the segments; segments subequal, oblanceolate, obovate or elliptical, obtuse, acute or bilobed at the tips, violet, violet-blue, lilac or sometimes white, often with a yellow throat. *Filaments* inserted near the top of the perianth tube, free, erect; anthers linear. *Ovary* small, subterranean; style filiform, erect, with three short stigmatic branches developed opposite the anthers; stigmas 3 or rarely more, elongated, usually with spathulate or sometimes lacerated tips or sometimes multifid. *Capsule* clavate or turbinate, with a narrow sterile pseudo-pedicellate base, sometimes ellipsoid, hygrochastic or rarely xerochastic; seeds numerous, small, globose or angled, brown or black. *Chromosome no.* 2n=12, rarely 22.

A small endemic genus of eight species, chiefly from the Karoo and dryer regions of the Cape Province, extending from Clanwilliam to East London and also north of the Orange River.

The genus is clearly divided into two subgenera. The monotypic subgenus *Rhipidopsis* differs in the shape of its corm, leaf anatomy, multifid stigmas, xerochastic capsules and in chromosome number, and might even be given generic status. It stands near *Crocus* of the northern hemisphere, differing in leaf and corm structure and in chromosome number. The two taxa are probably the result of parallel evolution.

Hooker f. described the flowers of *Syringodea* as shortly pedicellate. It is the sterile base of the ovary which forms a pseudo-pedicel in subgenus *Syringodea*; this elongates after flowering to push the top of the capsule above the ground.

Flowering period generally March to June. The flowers close at night and when fully open in the middle of the day they are usually salver-shaped.

The name *Syringodea* is derived from the Greek *syrinx*, meaning pipe, which alludes to the long tubular perianth tube.

Key to Subgenera

1 Corm almost turbinate or obovoid, with a small basal ridge; leaves several or rarely one; stigmas undivided or sometimes slightly lacerated .... 1. Subgenus *Syringodea*

1 Corm laterally flattened, almost lens-shaped, with a wide vertical, fan-shaped ridge; leaf usually single; stigmas multifid .... 2. Subgenus *Rhipidopsis*

Key to Subgenus *Syringodea*

1 Foliage leaf single, sometimes apparently 2 present, but then from adjacent corms .... 7. *S. saxatilis*

1 Foliage leaves several from a corm:

2 Perianth segments bilobed at their tips..1. *S. pulchella*

2 Perianth segments obtuse or acute:

   3 Limb of the perianth up to 16 mm across, the segments 5—8 × c. 2 mm..............................4. *S. flanaganii*

   3 Limb of the perianth 20 mm or more across, rarely only 16—19 mm, the segments 10 mm or longer, and wider than 4 mm, rarely less:

      4 Leaves 3—6 mm wide...6. *S. derustensis*

      4 Leaves 0,5—1,5 (—2) mm wide:

         5 Flowers (25—) 30—60 mm long, sometimes overtopping the leaves; corms somewhat obovoid with a basal ridge 4—7 mm wide; often in dense clusters; from the western and south-western Cape... 5. *S. longituba*

         5 Flowers (50—) 60—100 m long, not overtopping the leaves; corms single or few together, turbinate, with a very small ridge on a pointed base; from the eastern and north-eastern parts of the R.S.A.:

            6 Throat of perianth yellow or orange-yellow... 3. *S. bifucata*

            6 Throat of perianth not yellow or orange-yellow... 2. *S. concolor*

Subgenus *Rhipidopsis*

Only one species...8. *S. unifolia*

1. Subgenus **Syringodea**

Corm almost turbinate or obovoid, with a small ridge at the base. *Leaves* several or rarely one, without subepidermal collenchyma, with scattered tanniniferous idioblasts. *Style* erect; stigmas 3 with spathulate or sometimes lacerated tips. *Capsule* clavate or turbinate, wrinkled, with a narrow sterile pseudo-pedicellate base, hygrochastic with 6 valves. *Chromosome no.* 2n=12.

When moistened the ripe capsule loses its transverse wrinkles and elongates for about 20 per cent before it dehisces.

1. **Syringodea pulchella** *Hook. f.* in Curtis's bot. Mag. 29: t. 6072 (1873); Bak. in J. Bot., Lond. 5: 67 (1876); in J. Linn. Soc. Bot. 16: 85 (1877); Handb. Irid. 95 (1892); in F.C. 6: 34 (1896); Klatt in Abh. Naturforsch. Ges. Halle 15: 403 (1882); in Dur. & Schinz, Consp. Fl. Afr: 5: 160 (1895); Benth. & Hook. f., Gen. Pl. 3: 693 (1883); D(arnell) in Gdnrs' Chron. 81:79, fig. 41 (1927); De Vos in Jl S. Afr. Bot. 40: 229, fig. 11 (1974). Type: Cape, Hanover, in campis inter montes Sneeuberg, *Bolus* 1852 (K, holo.!; BOL!).

Plants 120—200 mm long. *Corm* with a very small pointed basal ridge. *Leaves* 3—6, filiform, with an adaxial groove, 100—125 × 0,8—1,5 (—2) mm, with a wider membranous sheath. *Bracts* reaching about halfway up the perianth tube. *Flowers* 1—4 (—5), 70—120 mm long, the limb 25—50 mm across, lilac, sometimes with a diffuse bluish blotch at the base of each segment. *Perianth tube* 60—100 mm long, c. 1,5 mm in diam., gradually widened upwards to c. 3 mm diam.; segments narrowly oblanceolate or elliptic-cuneate, 12—22 × 5—7 mm, emarginate or bilobed at the top, the lobes rounded, the outer segments purple-veined on the backs. *Stamens* exserted; filaments 3—4 mm long, white; anthers 5—9 mm long. *Style* 65—120 mm, white; stigmas c. 4 mm long, with spathulate tips, reaching halfway up, or above, the anthers. *Capsule* clavate, 20—30 mm long. Fig. 1:1.

Found in eastern districts of the Great Karoo between Middelburg and Graaff-Reinet and on the Middelburg and Somerset East mountain plateaux (3124—CB, DC; 3225—DA).

FIG. 1.—1, **Syringodea pulchella**, habit, × 1; la, perianth tube, stamens and pistil; 1b, capsules, closed and dehisced *(De Vos 2257 B)*. 2, **S. bifucata**, habit, × 1; 2a, outer (left) and inner (right) bract; 2b, transverse section of leaf *(Barker 10642 b)*. (All figures in this fascicle reproduced from the Jl S. Afr. Bot. with the Editor's permission).

1a
1
1b
2a
2
2b
5 mm
0,5 mm

Vouchers: *MacOwan* 827 (SAM); *Bolus* 1852 (BOL; K); *De Vos* 2257B (STE).

Flowering period March to April. Readily distinguished by its long-tubed lilac flowers with perianth segments bilobed at the top.

2. **Syringodea concolor** *(Bak.) De Vos* in Jl S. Afr. Bot. 40: 233, fig. 12 (1974).

S. bicolor Bak. var. *concolor* Bak., Handb. Irid. 96 (1892); in F.C. 6: 35 (1896). Type: Cape, Coetzierskraal, Murraysburg, *Tyson* 346 (K, holo.!; BOL!; SAM!).

Closely related to *S. pulchella* (no. 1) from which it differs as follows: *Bracts* 25—35 mm. *Flowers* 60—100 mm long, pale violet, bluish-violet, lilac or sometimes almost white, darker violet in the throat. *Perianth tube* (40—) 50—80 mm long; segments oblanceolate to elliptical, often slightly concave, subacute to obtuse, often slightly wider. *Filaments* 5—8 mm long, equal to the anthers. *Style* 50—85 mm long; stigmas with slightly spathulate or lacerated tips.

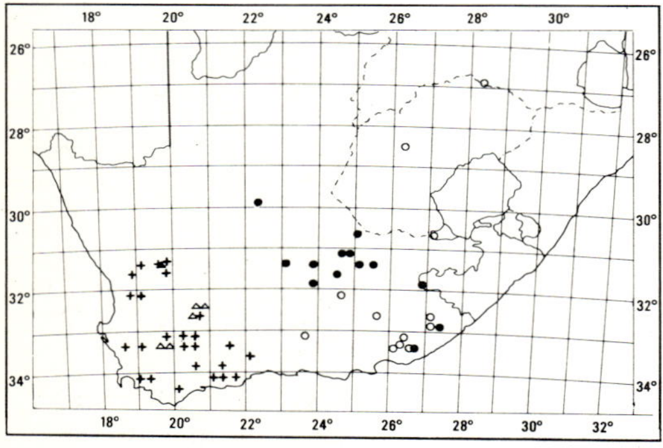

MAP 1.—● **Syringodea concolor**
○ **S. bifucata**
+ **S. longituba**
△ **S. unifolia**

Widely distributed in the northern, central and eastern parts of the Great Karoo from Prieska, Victoria West and Colesberg to Queenstown and King William's Town and also near Grahamstown. Map 1.

Vouchers: *Acocks* 14298, 17973; *Malan* in STE 30369; *Thorns* s.n., Apr. 1944 (NBG); *Bryant* J301 (BOL; PRE; K).

Distinguished by its large corm with a somewhat pointed base, large long-tubed, pale violet or pale blue to almost white flowers which are not bilobed at the apices of the perianth segments and are without a yellow throat. The seeds are black with a fine reticulate-foveate surface.

3. **Syringodea bifucata** *De Vos,* nom. nov.

S. bicolor Bak. in J. Bot., Lond. 5: 67 (1876), partly as to descr. and specimens cited but not as to syn., nom. illeg. (nom. superfl.); in J. Linn. Soc., Bot. 16: 86 (1877); Handb. Irid. 96 (1892); in F. C. 6: 35 excl. syn. and var.; sensu Klatt in Abh. Naturforsch. Ges. Halle 15: 403 (1882); sensu De Vos in Jl S. Afr. Bot. 40: 235, fig. 13 (1974). Lectotype: Orange Free State, Fat River, *Burke* 446 (K!).

Plants 60—200 mm long. *Corm* with a very small ridge on a pointed base. *Leaves* several, filiform with an adaxial groove, 60—200 × 0,5—1 (—1,5) mm, finely ciliate. *Bracts* reaching above to below the middle of the perianth tube. *Flowers* (50—) 65—100 mm long, the limb (25—) 40—55 mm across, pale to bright violet or sometimes white, with an orange-yellow throat. *Perianth tube* (30—) 45—70 mm long, widened at the top to 3 mm in diam.; segments obovate to oblanceolate, often slightly concave, (12—) 18—28 mm × (5—) 7—10 mm, obtuse or subobtuse. *Filaments* 5—8 mm, yellow, anthers 5—8 mm long. *Style* 50—75 mm, pale yellow; stigmas linear, reaching more or less the anther tips. *Capsule* clavate, 15—20 mm long; seeds brown with a rather coarse reticulate-foveate testa. Fig. 1:2.

Widely distributed in northern and eastern districts of the Great Karoo from Colesberg to Stutterheim and Willowmore, also near Grahamstown and north of the Orange River in the Orange Free State and at Vaaldam, Transvaal, Map. 1.

Vouchers: *Cheadle* 751; *Dyer* 1312 (GRA); *Zeyher* 446 (SAM; K); *Cruden* 256 (GRA; STE); *Davidson* 3108 (PRE; STE).

Distinguished by its large corm with a somewhat pointed base and large, long-tubed, violet flowers with a yellow throat, filaments and style. Variation occurs in the width of the perianth segments.

Baker (1876) cited *Trichonema longitubum* Klatt as a synonym when he described *S. bicolor* and Kuntze, Rev. Gen. 3: 309 (1898), on transferring the former to *Syringodea*, relegated *S. bicolor* to synonymy. Two species are involved, as the five collections cited by Baker under *S. bicolor* differ from the holotype of *S. longituba* found in B. They are without a legitimate name and a new name, *S. bifucata*, is therefore given which refers, like the illegitimate epithet *bicolor*, to the two-coloured perianth.

4. **Syringodea flanaganii** *Bak.* in Kew Bull. 1893: 158 (1893); in F.C. 6: 35 (1896); De Vos in Jl S. Afr. Bot. 40: 237 (1974). Type: Cape, top of Gonubie Hill near Draaibos and Komga, *Flanagan* 720 (K, holo.!; BOL!; PRE!; SAM!; B!).

Closely related to *S. bifucata* (no. 3), differing mainly in the smaller size of all the organs, e.g.: Plants 30−120 mm long. *Leaves* 30−100 × 0,5−1 mm. *Flowers* (25−) 30−50 mm long, limb up to 16 mm across, violet with a yellow throat. *Perianth tube* c. 25−45 mm long, widened at the top to 1,5 mm in diam.; segments oblanceolate or oblanceolate-elliptical, 5−8 × c. 2 mm. *Filaments* c. 2 mm; anthers c. 3 mm long. *Style* as long as the perianth tube or slightly longer; stigmas c. 1 mm long, spathulate. *Capsule* 7−8 mm long.

From the eastern Cape Province in the Port Elizabeth and Stutterheim districts (3227−DB; 3225−DC).

Vouchers: *Long* 989 (BOL; GRA; PRE; K); *Reed* s.n. (GRA); *Long* s.n. (BOL); *Bayliss* 2760 (NBG); *Paterson* 2452 (BOL; GRA).

May perhaps be regarded as a smaller variety of *S. bifucata* (no. 3), but as it is readily distinguishable from the latter, it is here treated as a distinct species.

5. **Syringodea longituba** *(Klatt)* Kuntze, Rev. Gen. 3: 309 (1898), excl. syn.; De Vos in Jl S. Afr. Bot. 40: 239 (1974). Type: Cape, without precise locality, *Mund & Maire* 1101 (B, holo.!).

Plants 40−100 mm long. *Corm* almost obovoid, oblique at the base, with a basal ridge 4−6 mm long. *Leaves* 5−8, filiform or somewhat flattened, with an adaxial groove, 15−60 × 0,8−1,5 (−2) mm, curved or flexuose, sometimes with a clockwise twist. *Bracts* 15−25 mm long, usually with fine, short brown lines. *Flowers* (25−) 30−50 mm long, the limb (16−) 20−25 (−30) mm across, violet or violet-blue, the throat yellow, orange-yellow, white or violet. *Perianth tube* (15−) 20−33 mm long, c. 1 mm in diam., widened to 3 mm at the top; segments obovate, slightly concave, (8−) 10−15 × (3−) 4−5 mm, obtuse to subacute. *Filaments* 3−4 (−8) mm, white, yellow or sometimes purple; anthers exserted, about as long as the filaments. *Style* (20) 25−40 mm; stigmas with the tips sometimes slightly widened. *Capsule* turbinate, 8−25 mm long.

Found in the south-western part of the Cape Province. Map 1.

A very variable species which differs from the preceding species in its corm with an obliquely flattened, not pointed, base with a wider basal ridge, in its shorter leaves which usually do not overtop the flowers, and in its more western distribution. Two varieties are recognized (see descriptions below for distinguishing characters):

(a) var. **longituba.**

Trichonema longitubum Klatt in Linnaea 34: 665 (1865−66); Bak. in J. Bot., Lond. 5: 67 (1876); in F.C. 6: 35, pro syn.

Syringodea filifolia Bak. in J. Bot., Lond. 5: 67 (1876); in J. Linn. Soc., Bot 16: 86 (1877); Handb. Irid. 96 (1892); in F.C. 6: 35 (1896); Klatt in Abh. Naturforsch. Ges. Halle 15: 403 (1882); in Dur. & Schinz, Consp. Fl. Afr. 5: 160 (1895); Lewis in Flower. Pl. Afr. 14:t. 547 (1934). Type: Cape, without precise locality, *Bowie* s.n. (BM. holo.!).

S. montana Klatt in Abh. Naturforsch. Ges. Halle 15: 403 (1882); in Dur. & Schinz, Consp. Fl. Afr. 5:160 (1895); Bak., Handb. Irid. 95 (1892); in F.C. 6: 34 (1896). Type: Cape, Hantamberge, *Meyer* s.n., anno 1869 (B, holo.!).

S. leipoldtii L. Bol. in J. Bot., Lond. 69: 12 (1931); Stopp in Bot., Stud., Jena 8: 38 (1958). Type: Cape, near Malmesbury, *Leipoldt* in BOL 19124 (BOL; holo.!; K!).

S. marlothii Schltr., ined. (B; PRE).

Perianth with an orange-yellow or yellow throat. Filaments, style and stigmas yellow.

Found in the western parts of the Great Karoo, western and south-western Cape districts eastwards to Riversdale.

Vouchers: *Salter* 2400 (BOL); 5370 (BOL; BM); *Leipoldt* in BOL 19124, 20326 (BOL); *De Vos* 2268 STE).

(b) var. **violacea** De Vos in Jl S. Afr. Bot. 40: 243 (1974). Type: Cape, Ladismith, 19 km NE of Muiskraal, *De Vos* 2270 (STE).

Perianth with a violet or white throat, sometimes with a violet median vein running downwards from each segment. Filaments, style and stigmas white. Fig. 2:1.

From the Little Karoo with outliers to the Laingsburg district and towards the southern coastal districts of Bredasdorp and Riversdale.

Vouchers: *Esterhuysen* 17112 (BOL); *Marloth* 9925; *De Vos* 2266 (STE).

6. **Syringodea derustensis** *De Vos* in Jl S. Afr. Bot. 40: 245, fig. 15 (1974). Type: Cape, Oudtshoorn, farm Drinkrivier near De Rust, *De Vos* 2269 (STE, holo.!).

Plants 50−80 mm long. *Corm* with an oblique basal ridge c. 5 mm long. *Leaves* 3−5, lanceolate or lanceolate-linear, sometimes conduplicate, curved, spreading, sub-

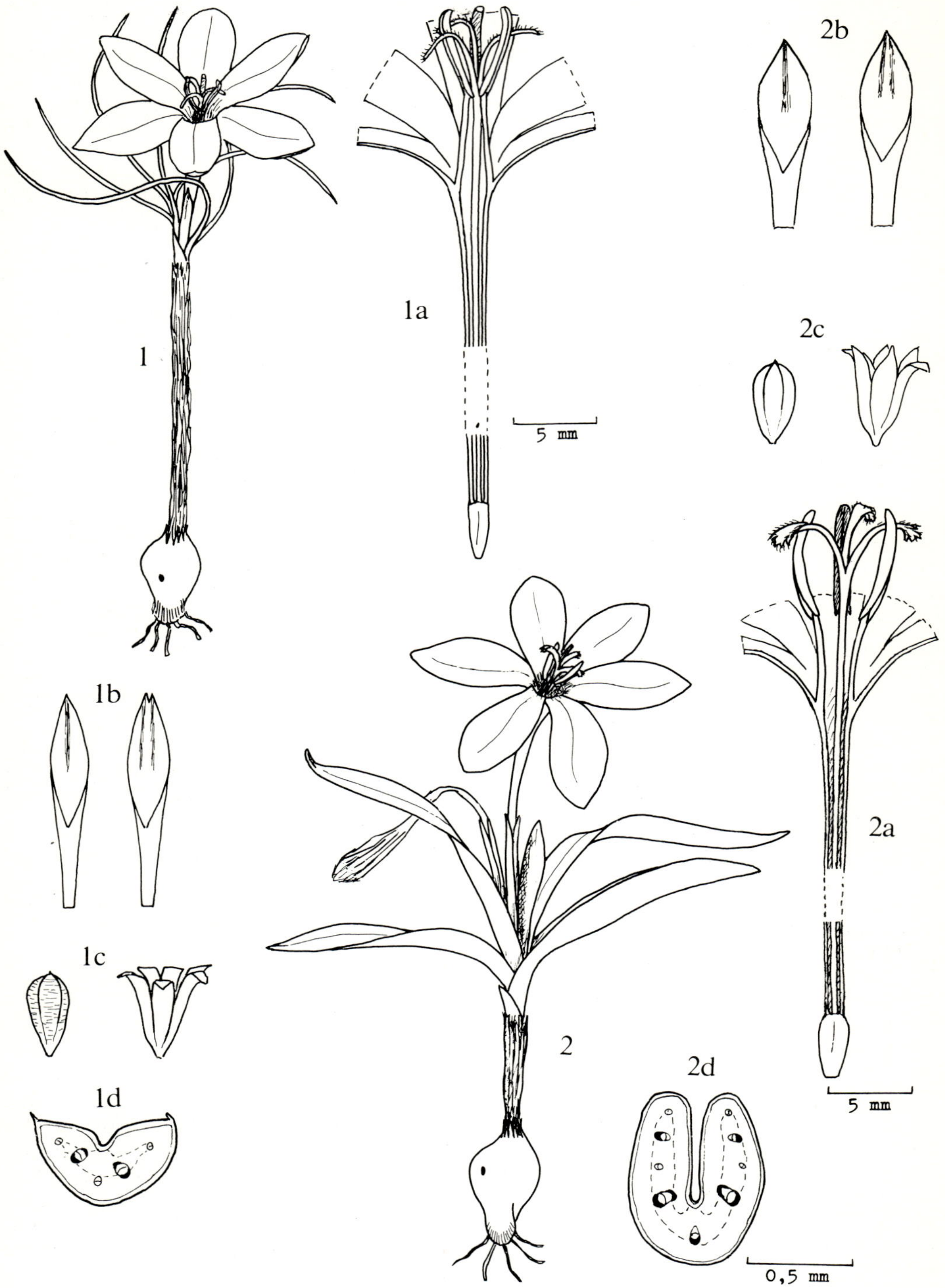

obtuse, 30−60 × 3−6 mm, often slightly swollen. *Bracts* reaching above the middle of the perianth tube. *Flowers* 45−60 mm long, the limb 30−40 mm across, pale violet or sometimes white. *Perianth tube* 25−35 mm long, widened at the top to 2 mm in diam.; segments obovate-cuneate, slightly concave, 15−20 × 8−10 mm, obtuse or subobtuse, the inner segments slightly wider than the outer. *Filaments* 5−6 mm, white; anthers 5−7 mm long. *Style* 33−38 mm, white; stigmas with spathulate or slightly lacerated tips. *Capsule* turbinate, ellipsoid or sometimes subglobose, 5−10 mm long. Fig. 2:2.

From the Little Karoo, on a stony koppie near De Rust, Oudtshoorn district (3322−BC), where it is common.

Vouchers: *Dahlstrand 2056*; *De Vos 2269* (STE).

Related to *S. longituba* (no. 5) and differs mainly in its wider, lanceolate or lanceolate-linear, strongly curved, spreading leaves.

7. **Syringodea saxatilis** *De Vos* in Jl S. Afr. Bot. 40: 246 (1974). Type: Cape, Ladismith commonage, c. 300 m from Winery, *De Vos 2264* (STE, holo.!).

Plants 40−80 mm long. *Corm* with an oblique basal ridge c. 5−7 mm long. *Leaf* single or sometimes apparently 2, strongly curved, spreading, 50−120 × 1,5−2,5 mm, lower half with an adaxial, finely ciliate groove, upper half terete. *Bracts* reaching below the middle or almost to the top of the perianth tube. *Flowers* 25−35 mm long, the limb 20−40 mm across, rose-lilac to pale violet. *Perianth tube* 15−20 mm long, widened at the top to 3 mm in diam.; segments obovate-lanceolate, obtuse to subacute, 10−20 × 5−8 mm. *Filaments* 5−7 mm long, white; anthers 3−5 mm long. *Style* 20−25 mm, white; stigmas with spathulate tips, reaching above or below the anther tips. *Capsule* turbinate. Fig. 3:1.

From the Little Karoo, on a stony koppie near Ladismith (3321−AD) where it is common.

Vouchers: *Stayner s.n.*, 30−5−71 (NBG); *De Vos 2264* (STE).

Related to *S. longituba* (no. 5) and *S. derustensis* (no. 6) and differs mainly in its single, strongly curved leaf which is unifacial and terete in its upper half and ciliate on the margins in the lower. Young plants which have not yet flowered have the whole leaf unifacial and terete, and not ciliate.

2. Subgenus **Rhipidopsis**

Rhipidopsis *De Vos* in Jl S. Afr. Bot. 40: 249 (1974). Type species: *S. unifolia* Goldbl.

Corm flattened laterally and almost lens-shaped, with a wide fan-shaped ridge separating the two faces. *Leaf* usually single, slightly swollen, with subepidermal collenchyma in the adaxial groove and without tannin. *Style* later bent; stigmas multifid. *Capsule* ellipsoid, xerochastic with three valves. *Chromosome no.* 2n=22.

The name *Rhipidopsis* refers to the fan-shaped corm tunics.

8. **Syringodea unifolia** *Goldbl.* in Flower. Pl. Afr. 41: t. 1638 (1971); De Vos in Jl S. Afr. Bot. 40: 249 (1974). Type: Cape, Worcester, Matroosberg opposite hut, *Stayner* in NBG 87602 (NBG holo.; BOL 30678!).

S. rosea sensu Klatt in Abh. Naturforsch. Ges. Halle 15: 403 (1882), excl. syn. *Ixia rosea* L.; Bak. in F. C. 6: 35 (1896).

Plants 50−120(−200) mm long. *Corm* laterally flattened, almost lens-shaped, with a wide fan-shaped vertical ridge, 10−20 (−25) mm wide. *Foliage leaf* single, rarely 2, usually falcate, 50−150 × 1,5−3,5 mm, swollen, with a wide adaxial groove, the upper part often terete, *Peduncle* short, elongating to 25 mm in the fruiting stage. *Bracts* membranous, reaching above the middle or almost to the top of the perianth tube. *Flowers* 1−4, somewhat salver-shaped, 40−60 mm long, the limb 30−65 mm across, pale violet, violet-blue or rarely white, the throat often orange-yellow.

FIG. 2. −1, **Syringodea longituba** var. **violacea**, habit, × 1; 1a, perianth tube, stamens and pistil; 1b, outer (left) and inner (right) bract; 1c, capsules, closed and dehisced; 1d, transverse section of leaf (*De Vos 2270*). 2, **S. derustensis**, habit, × 1; 2a, perianth tube, stamens and pistil; 2b, outer (left) and inner (right) bract; 2c, capsules, closed and dehisced; 2d, transverse section of a folded leaf (*De Vos 2269*).

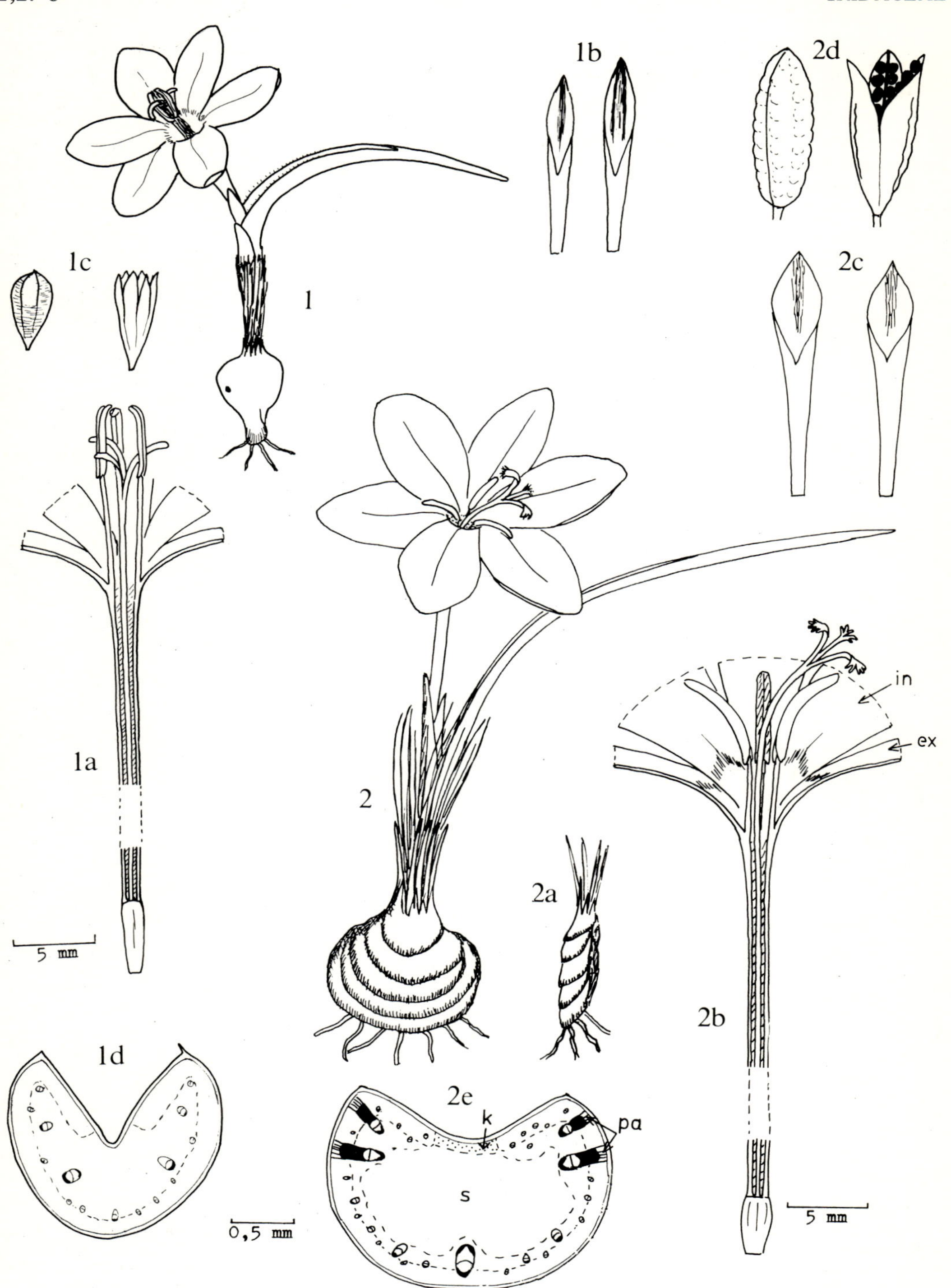
1b
2d
1c
2c
1
1a
5 mm
2
2a
2b
in
ex
1d
2e
k
pa
s
0,5 mm
5 mm

Perianth tube 30−40(−50) mm long, c. 1 mm in diam., widened to c. 3 mm at the top; segments elliptic or obovate-cuneate, somewhat concave, 15−30 × 7−14 mm, obtuse. *Stamens* exserted; anthers 5−10 mm long, slightly spreading, orange. *Style* 45−65 mm, curved at the top; stigmatic branches multifid and lacerate. *Capsule* trigonous-ellipsoid. Fig. 3:2.

Found in the Sutherland, Calvinia and Worcester districts on mountains and mountain plateaux. Map 1.

Vouchers: *Stayner* s.n., 19.7.68 (NBG); *De Vos* 2217, 2267 (STE); *Marloth* 9799; *Stayner* in NBG 87594, 90073.

Readily distinguished by its flattened fan-shaped · corm, usually single, rather swollen, subterete leaf, and long-tubed flowers with numerous stigmas and a curved style.

Excluded species

Syringodea latifolia Klatt in Abh. Naturforsch. Ges. Halle 15: 403 (1882) is **Hesperantha latifolia** (Klatt) De Vos in Jl S. Afr. Bot. 40: 252 (1974), non Steud. (1841), nom. nud.

Syringodea linifolia Phill. in Ann. S. Afr. Mus. 9: 125 (1913) is **Duthiastrum linifolium** De Vos in Jl. S. Afr. Bot. 41: 91 (1975).

Syringodea luteo-nigra Bak. in Kew Bull. 1897: 281 is **Romulea macowanii** Bak. in J. Bot., Lond. 5: 236 (1876).

Syringodea minuta Klatt in Abh. Naturforsch. Ges. Halle 15: 403 (1882) is **Pauridia minuta** (L.f.) Dur. & Schinz, Consp. Fl. Afr. 5: 142 (1895).

FIG. 3.−1, **Syringodea saxatilis**, habit, × 1; 1a, perianth tube, stamens and pistil; 1b, outer (left) and inner (right) bract; 1c, capsules, closed and dehisced; 1d, transverse section of leaf *(De Vos)*. 2, **S. unifolia**, habit, × 1; 2a, corm, lateral view; 2b, perianth tube, stamens and pistil; 2c, outer (left) and inner (right) bract; 2d, capsules, closed and dehisced; 2e, transverse section of leaf: k, collenchyma; pa, parenchyma; s, space *(De Vos* 2267).

Romulea *Maratti,* Fl. Rom. Sat. 13 (1772), nom. conserv.; Seb. & Mauri, Fl. Rom. Prodr. 11 (1818); Bak. in J. Linn. Soc., Bot. 16: 86 (1877); Handb. Irid. 97 (1892); in F.C. 6: 36 (1896); in F.T.A. 7: 344 (1898); Klatt in Abh. Naturforsch. Ges. Halle 15: 398 (1882); Benth. & Hook. f., Gen. Pl. 3: 694 (1883); Pax in Natürl. PflFam. 2, 5: 143 (1888); Baill., Hist. Pl. 156 (1894); Béguinot in Bot. Jb. 38: 322 (1907); in Malpighia 21: 49 (1907); 22: 377 (1908); 23: 55 (1909); Marloth, Fl. S. Afr. 4: 147 (1915); Diels in Natürl. PflFam. edn 2, 15a: 474 (1930); Lewis in Adams. & Salter, Fl. Cape Penins. 220 (1950); Phill., Gen. edn 2, 212 (1951); De Vos in Jl S. Afr. Bot. Suppl. 9: 1 (1972); Goldbl. in R.A. Dyer, Gen. 965 (1976). Type species: *R. bulbocodium* (L.) Seb. & Mauri (= *Crocus bulbocodium* L.) (typ. cons.).

Crocus L., Sp. Pl. 1: 36 (1753), partly; Gen. Pl. edn 5, 23 (1754), partly.

Bulbocodium Mill., Fig. Pl. 160 (1760); Kuntze, Rev. Gen. 2: 700 (1891); non L.

Ixia L., Sp. Pl. edn 2, 51 (1762), partly; Roem. & Schult., Syst. Veg. 1: 373 (1817).

Ilmu Adans., Fam. Pl. 2: 497 (1763) nom. prius, nom. rej.

Trichonema Ker-Gawl. in Curtis's bot. Mag. 16: t. 575 (1802); in Kön. & J. Sims, Ann. Bot. 1: 222 (1805); Irid. Gen. 79 (1827); Ait., Hort. Kew. edn 2, 1: 82 (1810); Gray, Nat. Arr. Brit. Pl. 2: 195 (1821) as *Trichomema;* Spreng., Gen. Pl. 1: 37 (1830); Klatt in Linnaea 34: 659 (1865—66); Harv., Gen. Pl. 330 (1838), edn 2, 376 (1868). Type species: *T. cruciatum* (Jacq.) Ker-Gawl. (*Ixia cruciata* Jacq.).

Spatalanthus Sweet, Brit. Fl. Gdn 3: t. 300 (1829). Type species: *S. speciosus* Sweet.

Plants small, deciduous. *Corm* small, subglobose, (ob)ovoid, bell-shaped or asymmetric, with a rounded, pointed, or flattened base and often a circular or crescent-shaped basal ridge (Figs 7: 1, 4: 1 & 2, 11: 2); tunics almost entire, woody, split at base into small fibrils or pointed teeth, and at top into long pointed teeth. *Stem* short, hidden by leaf bases, or elongated, erect, extending above ground. *Cataphylls* 1—3, sheathing base of shoot. *Foliage leaves* several, all basal or 1—3 basal and some cauline, with a bifacial sheath and a unifacial, filiform or compressed cylindrical, usually 4-grooved, rarely 2-grooved or up to 8-grooved or 4-winged blade. *Inflorescence* with 1 or more flowers, each terminal on a suberect peduncle, sessile, enclosed at the base by a 2-valved spathe. *Spathe valves (bracts)* green or submembranous, inner with wide membranous margins or rarely wholly membranous. *Flowers* actinomorphic, funnel-shaped or bell-shaped, rarely salver-shaped, of various colours, thermonastic. *Perianth tube* short, usually funnel-shaped, rarely long; segments equal or subequal, usually oblanceolate or elliptical, acute to obtuse. *Filaments* usually inserted near base of perianth tube, erect, usually free; anthers linear. *Ovary* small; style filiform, erect, with 3 short, deeply bifid, rarely multifid branches developed opposite the stamens; stigmas 6, rarely more, grooved above. *Capsules* shortly cylindrical to subglobose, loculicidally 3-valved; seeds numerous, small, globose or angled, brown.

An African and Mediterranean genus of about 90 species, with outliers to the Canaries and to Britain (one species), and with a large developmental centre in the Cape Province where 69 species occur, several with a very restricted distribution.

Flowering period generally late winter to late spring. The flowers usually open around noon and close in the late afternoon. The unifacial leaf blades are based on a general plan of four ribs and four stomatiferous grooves. Variation from this is comparatively rare, e.g. 5—8-grooved leaves occur in *R. aquatica* (no. 27), X-shaped ones in *R. hirta* (no. 44), and leaves with the adaxial groove open almost to the leaf tip in *R. tortuosa* (no. 41).

Key to Subgenera

1 Flowers funnel-shaped or bell-shaped, of various colours; perianth segments ascending, often becoming reflexed; perianth tube short or rarely long .... 1. Subgenus ROMULEA (below)

1 Flowers salver-shaped, not yellow; perianth segments spreading at right angles from a long, narrow perianth tube .... 2. Subgenus LOMUREA (p. 2,2: 16)

Key to Sections of Subgenus *Romulea*

1 Basal foliage leaves 4-winged, X-shaped in transverse section, with two wide lateral grooves and a strong vein up the middle of each groove ....4. Section *Hirtae* (p. 2,2: 14)

1 Basal foliage leaves filiform or compressed cylindrical, with 4 grooves and 4 ribs, or rarely with 2 or 5−8 grooves and as many ribs, or with rib margins widened into 8 wings:

  2 Corm rounded or pointed at base or rarely with a very small, almost pointed basal ridge much narrower than diameter of corm (Figs 13: 1, 15), tunics split at base into coarse, bent or straight, acuminate teeth, or rarely more slender teeth or fibrils:

    3 Inner bract without two stronger veins and not distinctly 2-keeled; outer without a stronger median vein .... 5. Section *Roseae* (p. 2,2: 15)

    3 Inner bract usually with two stronger veins especially in upper half, 2-keeled; outer bract often with a stronger median vein in upper half ....6. Section *Spatalanthus* (p. 2,2: 16)

  2 Corm bell-shaped with a circular basal ridge or obliquely flattened towards base with a crescent-shaped or fan-shaped basal ridge about as wide as corm or wider (Figs 4, 7: 1, 11: 2), tunics usually split on ridge into fine parallel fibrils or small fibril clusters:

    4 Basal ridge of corm often fan-shaped and almost vertical, wider than corm (Fig. 11: 2); flowers yellow, sometimes with a long perianth tube; bracts membranous or submembranous in lower half, green or greenish in upper, inner bracts with wide membranous margins .... 3. Section *Tortuosae* (p. 2,2: 14)

    4 Basal ridge of corm circular or crescent-shaped, horizontal or oblique, as wide as corm or slightly narrower (sometimes wider but bracts then green) (Figs 7/1 & 4); flowers of various colours, rarely with a long perianth tube but flowers then pink; bracts largely green or sometimes submembranous, inner generally green with wide, membranous margins or rarely wholly scarious:

      5 Corm tunics on basal ridge split into a fringe of very slender or minute, parallel or sometimes irregularly grouped fibrils, or rarely into a row of slender parallel, very sharply bent fibrils which later break off on ridge... 1. Section *Romulea* (below)

      5 Corm tunics on basal ridge split into a row of small, rounded or elongated clusters of minute fibrils, with a root extending from centre of some clusters .... 2. Section *Aggregatae* (p. 2,2: 14)

1. Key to Section *Romulea*

1 Corm symmetrical, bell-shaped, with a circular basal ridge around a flat or slightly concave basal disc:

  2 Leaves spirally twisted .... 25. *R. tortilis*

  2 Leaves straight or bent but not spirally twisted:

    3 Flowers white or cream inside, with a yellow cup, without dark blotches in the throat:

      4 Outer perianth segments greenish or cream on the backs; basal ridge of corm not much wider than the corm itself, with tunics usually split into regularly arranged parallel fibrils; peduncles of fruiting specimens suberect or slightly bent ....22. *R. triflora*

      4 Outer perianth segments purplish or reddish on the backs, or sometimes greenish; basal ridge of corm much wider than the corm with tunics irregularly lacerated or split into irregular groups of fibrils on the ridge; peduncles of fruiting specimens very widely patent ....24. *R. sladenii*

    3 Flowers magenta, pink, yellow or apricot inside, often with a yellow cup and dark blotches in the throat:

      5 Perianth generally bright yellow, sometimes pale, rarely with brown blotches in the throat ....22. *R. triflora*

      5 Perianth magenta, pink or apricot, often with dark red or purplish black blotches in the throat:

        6 Segments of perianth usually more than 6 mm wide, rarely less and then apricot coloured; flowers deep rosy-pink, magenta or apricot inside, usually with dark blotches in the throat ....21. *R. hirsuta*

        6 Segments of perianth less than 5(−6) mm wide; flowers pale pink inside without dark blotches ....23. *R. gracillima*

1 Corm asymmetrical, obliquely flattened towards the base, with a wide or small crescent-shaped basal ridge:

  7 Plants aquatic, with the corm and base of shoot submerged; unifacial part of the basal foliage leaf 5−8-grooved and 5−8-ribbed; inner perianth segments wider than the outer and different in shape:

    8 Basal foliage leaves 2; perianth yellow; style more than 10 mm long .... 26. *R. multisulcata*

8 Basal foliage leaf single; perianth white or cream in the upper half, yellow in the lower; style less than
   10 mm long ....27. *R. aquatica*

7 Plants terrestrial, sometimes in marshy localities; unifacial part of the basal foliage leaves 4-grooved and
   4-ribbed, rarely 2-grooved or 3-angled; inner and outer perianth segments more or less similar:

  9 Anther connectives attenuate, produced 2,5−6 mm above the thecae ....4. *R. flexuosa*

  9 Anther connectives not as above, not or hardly produced above the thecae:

   10 Outer bract green with pronounced brown-streaked or brownish membranous tip and margins:

      Perianth yellow; outer bract without a stronger median vein ....19. *R. pearsonii*

      Perianth bright magenta; outer bract often with a stronger median vein ....20. *R. neglecta*

   10 Outer bract green or submembranous, with membranous margins, if present, very narrow:

     11 Flowers yellow, or rarely orange:

      12 Filaments at least twice as long as the anthers ....,...10. *R. sulphurea*

      12 Filaments not or only slightly longer than the anthers:

        13 Inner bract in the median zone and up to the tip as green as the outer bract:

          14 Membranous margins of the inner bract brownish, brown-edged or minutely
             brown-streaked, especially in the upper half; perianth segments with dark blotches or
             dark lines in the throat, oblanceolate; basal ridge of the corm often wider than the
             corm, often with irregularly grouped basal fibrils ....11. *R. montana*

          14 Membranous margins of the inner bract colourless; perianth segments without dark
             blotches inside, the outer segments narrowly elliptical; basal ridge of the corm not
             wider than the corm, with minute parallel basal fibrils ....9. *R. elliptica*

        13 Inner bract wholly scarious or submembranous, or greenish in the median zone but not as
           green as the outer:

          15 Plant with two basal foliage leaves when the stem is elongated; inner bract
             submembranous in the median zone, with wide, scarious margins ....18. *R. citrina*

          15 Plant with a single basal foliage leaf when the stem is elongated; inner bract wholly
             scarious or rarely submembranous in the median zone, with wide membranous
             margins:

            16 Perianth sulphur-yellow or pale yellow; inner bract usually scarious ....15. *R. flava*

            16 Perianth bright golden-yellow or orange-yellow; inner bract often greenish in the
               upper half or with reddish veins ....16. *R. saldanhensis*

     11 Flowers not yellow except for a frequently yellow cup and rarely a yellow throat:

      17 Perianth tube more than 16 mm long, longer than the segments ....8. *R. kamisensis*

      17 Perianth tube rarely up to 12 mm long, shorter than the segments:

        18 Outer bract keeled, much longer than the inner bract and sometimes almost as long as the
           flower ....2. *R. papyracea*

        18 Outer bract not keeled, subequal to the inner bract or slightly longer, much shorter than
           the flower:

          19 Full-grown corm somewhat vertically elongated, higher than wide, with a high, almost
             chisel-shaped or sometimes slightly wavy basal ridge; flowers (7−)10−15(−20) mm
             long; membranous margins of inner bracts usually with relatively large brown spots:

            20 Flowers 7−15 mm long, pale mauve or rarely almost white, often with a darker mauve
               throat; perianth segments less than 3 mm wide; brown spots on the inner bract well
               defined ....28. *R. minutiflora*

            20 Flowers 15−20 mm long, cream or white; perianth segments 3−4 mm wide; brown
               spots on the inner bract faint or absent ....29. *R. sinispinosensis*

          19 Full-grown corm more or less isodiametric, with a low or sometimes high
             crescent-shaped basal ridge; flowers more than 16 mm long, rarely only 12 mm; inner
             bract with colourless, brownish or brown-streaked membranous margins or sometimes
             wholly membranous:

            21 Tunics of corm with a row of parallel fibrils or slender teeth very sharply bent over and
               ultimately broken on a small, rather high, crescent-shaped basal ridge; flowers
               (12−)16−20(−24) mm long:

22 Outer bract green in the upper half, submembranous in the lower; stem short, hidden by leaf bases .... 30. *R. pratensis*

22 Outer bract green; stem usually extended from the leaf bases.................. 31. *R. gigantea*

21 Tunics of corm with a row of fine parallel fibrils on a low, rather wide, basal ridge; flowers rarely less than 20 mm long:

23 Inner bract scarious or with a submembranous median zone; one or sometimes two basal foliage leaves present when the stem is elongated:

24 Flowers white, with large black blotches usually bordered by a yellow margin in the throat; leaves 2-grooved, with one wide and one narrow rib, appearing 3-angled .... 17. *R. barkerae*

24 Flowers white, cream or blue without blotches in the throat; leaves 4-grooved and 4-ribbed:

25 Cup of perianth yellow, segments white or rarely blue; a single basal foliage leaf present when the stem is elongated; inner bract usually wholly scarious .... 15. *R. flava*

25 Cup and lower third or half of perianth segments yellow, upper part white, cream, blue or rarely bluish violet; 1−2 basal foliage leaves present when the stem is elongated; inner bract submembranous in the median zone:

26 Upper part of perianth segments cream or white, 5−8 mm wide; outer segments not blotched on the backs .... 14. *R. leipoldtii*

26 Upper part of perianth segments lavender-blue with a pale transverse band when fresh, rarely bluish violet or white, up to 5 mm wide; outer segments often irregularly blotched on the backs .... 13. *R. tabularis*

23 Inner bract with a green median zone and wide membranous margins; two basal foliage leaves present when the stem is elongated:

27 Filaments partly or wholly red or reddish black, or rarely yellow, inserted halfway up the perianth tube; flowers with very little or no yellow at the base; from Namaqualand .... 7. *R. namaquensis*

27 Filaments orange-yellow to pale yellow, inserted near the base or in the lower half of the perianth tube; flowers usually with an orange or golden yellow cup, or cup sometimes pale yellow and flowers then small, or sometimes with only a yellow or orange perianth tube; south of Namaqualand:

28 Perianth segments white or cream inside:

29 Membranous margins of the inner bract brown-edged; basal ridge of the corm often wider than the corm, with fibrils irregularly grouped .... 12. *R. toximontana*

29 Membranous margins of the inner bract colourless; basal ridge of the corm about as wide as the corm or narrower, with slender parallel fibrils .... 1. *R. schlechteri*

28 Perianth segments magenta or pink to pale violet, sometimes with dark blotches in the throat:

30 Flowers deep old rose or bright pink, with large maroon or violet blotches in the throat .... 6. *R. biflora*

30 Flowers magenta-pink or pale violet, without dark blotches in the throat:

31 Perianth segments generally obtuse; outer segments shiny and wine-coloured on the backs, often becoming violet on drying .... 5. *R. vinacea*

31 Perianth segments acute to subobtuse; outer segments variously coloured and marked on the backs but not shiny or wine-coloured:

32 Stem generally less than 100 mm long, rigid; flowers lilac-pink or pale mauve; filaments pilose near their bases ......1. *R. schlechteri*

32 Stem often more than 100 mm long, rarely short, very slender; flowers magenta-pink with reddish veins in the throat; filaments minutely pilose almost to their tops .... 3. *R. saxatilis*

2. Key to Section *Aggregatae*

1 Anthers longer than the filaments; flowers bright carmine-red or deep rosy-pink, sometimes with dark blotches in the throat, cup not bright yellow; corm symmetrical with a circular basal ridge, or slightly asymmetrical with a horse-shoe shaped basal ridge:

  2 Corm with a circular basal ridge; flowers carmine-red or deep rosy-pink with large purplish-black blotches in the throat, the cup with slender dark lines .... 32. *R. amoena*

  2 Corm with a horse-shoe shaped basal ridge; flowers carmine without dark blotches or a differently coloured cup .... 33. *R. sanguinalis*

1 Anthers subequal to the filaments or shorter; flowers not carmine-red or deep rosy-pink, rarely with dark blotches in the throat and the cup then orange-yellow; corm asymmetrical with a crescent-shaped or rarely horse-shoe shaped basal ridge:

  3 Membranous margins of inner bracts usually colourless; anthers never joined at their tips; stem short or rarely elongated up to 120 mm:

    4 Flowers yellow or apricot-coloured sometimes with dark blotches in the throat.................... 34. *R. setifolia*

    4 Flowers magenta-pink with dark veins in the throat.. 39. *R. albomarginata*

  3 Membranous margins of inner bracts generally brown, brown-edged or speckled in the upper half (colourless in the lower); anthers joined at their tips in young flowers; stem usually elongated, extended from the leaf sheaths:

    5 Corm tunics fibrous or split into narrow segments, with a dense collar of fibres 20—80 mm long around the base of the shoot.. 36. *R. fibrosa*

    5 Corm tunics smooth and hard, with a collar of fibres or acuminate teeth up to 15 mm long around the base of the shoot:

      6 Flowers orange .... 37. *R. jugicola*

      6 Flowers not orange except sometimes for an orange-yellow cup:

        7 Basal foliage leaf of elongated stem single, rarely two in young plants and the first leaf then shorter than the second; lateral leaf ribs narrower than the median ribs .... 38. *R. dichotoma*

        7 Basal foliage leaves of elongated stem 2, the first leaf longer than the second; four leaf ribs about equal in width:

          8 Flowers cream, pale yellow, greenish yellow or pale apricot, often with dark apricot-coloured veins .... 35. *R. longipes*

          8 Flowers magenta to pink with diffuse violet-blue blotches in the throat.................... 36. *R. fibrosa*

3. Key to Section *Tortuosae*

1 Perianth tube funnel-shaped, less than 10 mm long, shorter than the segments:

  2 Foliage leaves several, spirally twisted, flexuose or sometimes bent, without adhering sand particles:

    3 Corm with a wide, more or less vertical, fan-shaped basal ridge which is much wider than the corm; bracts largely membranous or submembranous, greenish towards the tips only; leaves usually spirally twisted or flexuose .... 41. *R. tortuosa*

    3 Corm with a somewhat more horizontal or oblique, crescent-shaped basal ridge which is not much wider than the corm itself; bracts green, greenish or reddish, submembranous in the lower half, the inner bract with wide brown-edged or brown-speckled membranous margins; leaves bent or suberect .... 40. *R. austinii*

  2 Foliage leaf generally single, rarely 2, suberect or bent, with adhering sand particles .... 42. *R. sphaerocarpa*

1 Perianth tube narrowly tubular for most of its length, widened in the upper part, generally longer than 15 mm, and longer than, equal to, or rarely slightly shorter than the segments .... 43. *R. macowanii*

4. Key to Section *Hirtae*

1 Corm with tunics split into long acuminate teeth bent over a rounded base; perianth pale yellow, often with a pale reddish brown or greenish yellow transverse band on each segment .... 44. *R. hirta*

1 Corm with tunics often split into minute parallel fibrils on a crescent-shaped basal ridge; perianth violet-rose to lilac, rarely salmon-pink, with a violet blotch or transverse band on each segment in the throat .... 45. *R. tetragona*

5. Key to Section *Roseae*

1 Corm more or less pointed at the base, the tunics split into almost straight basal teeth or fibrils converging to the basal point:

  2 Perianth yellow ....55. *R. membranacea*

  2 Perianth not yellow except for a sometimes yellow cup:

    3 Inner bract with colourless membranous margins; corm with slender teeth or fibrils on a very small ridge at the narrow, almost pointed base:

      4 Stamens and style not reaching halfway up the perianth; membranous margins of the inner bract wider than those of the outer .... 46. *R. autumnalis*

      4 Stamens and style reaching more than halfway up the perianth; membranous margins of the bracts subequal in width .... 47. *R. campanuloides*

    3 Inner bract with brown, brown-speckled or -streaked membranous margins; corm with coarse acuminate teeth converging to a basal point:

      5 Flowers (35−)40−60 mm long, old-rose with dark red blotches in the throat, the cup pale yellow or greenish yellow ....61. *R. eximia*

      5 Flowers 25−35 (−40) mm long, magenta to lilac-pink, often with a blue, violet or blue-black blotch or zone in the throat, the cup golden-yellow or orange-yellow:

        5a Bracts with inconspicuous veining .... 60. *R. cruciata*

        5a Bracts with strong, closely spaced, conspicuous veining ....48a. *R. vlokii*

1 Corm rounded at the base with the tunics split into bent acuminate teeth curved over the base of the corm:

  6 Outer bract with wide, pronounced, membranous margins and a large membranous tip:

    7 Perianth yellow, with or without dark blotches in the throat:

      8 Inner perianth segments about 2−4 mm wider than the outer; stigmas overtopping the anthers ....54. *R. diversiformis*

      8 Inner perianth segments subequal to the outer; stigmas not overtopping the anthers:

        9 Flowers more than 25 mm long, bright yellow, often with dark blotches in the throat; bracts firm, green, with brown-streaked membranous margins and tip ....49. *R. luteoflora*

        9 Flowers 15−25(−30) mm long, pale yellow, without dark blotches; bracts largely membranous or green in the centre of the upper half ....53. *R. malaniae*

    7 Perianth magenta, pink, lilac, white or rarely pale blue, with, or sometimes without dark blotches in the throat, cup variously coloured:

      10 Style branches multifid; stigmas small, terminal, 12 or more ....52. *R. multifida*

      10 Style branches usually bifid; stigmas elongated, usually 6 (except in *R. komsbergensis* which has rarely 7−10 stigmas):

        11 Perianth cup yellow, brown at its base when fresh; pollen brown or rust-coloured; anthers generally circinnate or incurved, not joined at their tips; bracts often submembranous in the lower half, green in the median upper half ....51. *R. komsbergensis*

        11 Perianth cup yellow or orange-yellow, not brown at its base; often with dark longitudinal stripes; pollen yellow; anthers erect or slightly incurved, at first joined at their tips; bracts with a firm green median zone and wide membranous margins:

          12 Outer bract with a linear green median zone; flowers magenta, pink, lilac or white, with or without dark blotches in the throat ....48. *R. atrandra*

          12 Outer bracts with a triangular green lower half; flowers pale blue with a violet and below that an almost black blotch in the middle of each segment ....50. *R. hallii*

  6 Outer bract with narrow, hardly visible membranous margins, the tip minutely or hardly membranous:

    13 Perianth segments magenta to lilac-pink, rosy-pink or white inside, without yellow tints except in the cup; peduncles straightening and becoming suberect on drying out in the fruiting stage ....59. *R. rosea*

    13 Perianth segments apricot, terra-cotta, old-rose, yellow or rarely white inside; peduncles of mature capsules on drying out, usually bending from their bases and widely patent (except in *R. monticola* which has suberect peduncles and yellow flowers, and perhaps also in *R. cedarbergensis*):

      14 Leaves 1−2(−3), ca. 0,5 or less in diam.; flowers 1 or rarely 2, white or very pale pink ....58. *R. cedarbergensis*

14 Leaves 3 or more, 0,5—1,5 mm in diam.; flowers 2 or usually more, not white:

    15 Flowers bright golden-yellow; anthers in young flowers joined at their tips; peduncles suberect or slightly curved in the fruiting stage when drying out ....57. *R. monticola*

    15 Flowers apricot, terra-cotta, deep old-rose or sometimes yellow; anthers never joined at their tips; peduncles bending from their bases and widely patent when mature capsules dry out .... 56. *R. obscura*

6. Key to Section *Spatalanthus*

1 Filaments joined into a short stout column ....65. *R. monadelpha*

1 Filaments free:

  2 Flowers yellow with dark blotches in the throat...63. *R. viridibracteata*

  2 Flowers red or bright pink with dark blotches in the throat:

    3 Leaves about 1 mm in diameter or less, filiform, with 4 narrow grooves .... 64. *R. sabulosa*

    3 Leaves 2—5 mm in diameter, somewhat swollen, 4-sided, 8-angled or 8-winged, with 4 wide grooves:

      4 Flowers bright pink with a purplish black blotch on each segment in the throat, cup bright yellow with a dark longitudinal line extending from each blotch .... 62. *R. subfistulosa*

      4 Flowers pinkish red with a brownish black blotch bordered by a pale violet zone on each segment and below that an elongated yellow blotch .... 66. *R. vanzyliae*

Key to sections of Subgenus *Lomurea*

1 Corm with a rounded or pointed base; foliage leaves several, usually more than 1 mm in diam.; style more than 20 mm long .... 7. Section *Lomurea* (below)

1 Corm with a crescent-shaped basal ridge; leaves 1(—2), less than 1 mm in diam.; style 15—20 mm long .... 8. Section *Stellanthe* (below)

7. Key to Section *Lomurea*

1 Perianth tube less than 25 mm long; style 22—30 mm long; corm with a rounded base and bent basal teeth .... 67. *R. syringodeoflora*

1 Perianth tube more than 30 mm long; style more than 50 mm long; corm with a pointed base and almost straight basal teeth ....68. *R. hantamensis*

8. Section *Stellanthe*

Only one species ....69. *R. stellata*

1. Subgenus **Romulea**

Bak. in F.C. 6: 36 (1896); De Vos in Jl S. Afr. Bot. Suppl. 9: 58 (1972).

Section *Euromulea* Diels in Natürl. PflFam. edn 2, 15a: 474 (1930).

Flowers funnel-shaped or bell-shaped, of various colours. *Perianth tube* funnel-shaped or rarely saucer-shaped, short, rarely longer than the segments; segments oblanceolate or sometimes elliptical, often becoming reflexed, acute to obtuse. *Stamens* usually inserted near the base of the perianth tube; filaments minutely pilose towards the base. *Capsules* produced well above ground.

This subgenus comprises all the species of *Romulea* except three.

1. Section **Romulea**

De Vos in Jl. S. Afr. Bot. Suppl. 9: 58 (1972).

'Stirps' *Hirsutae* Bég. in Malpighia 23: 87 (1909), partly; 'Stirps' *Subluteae* Bég., l.c. 98; 'Stirps' *Bulbocodioides* Bég., l.c. 107, partly; Section *Pratenses* De Vos in Jl S. Afr. Bot. Suppl. 9: 196 (1972).

Corm with tunics split into a fringe of fine parallel fibrils on a crescent-shaped or circular basal ridge. *Stem* short, hidden by leaf bases, or elongated. *Foliage leaves* all basal or 1 − 2 basal and a few cauline, terete or compressed-cylindrical, with 4 or rarely 5 − 8 grooves. *Bracts* largely green, inner with wide membranous margins or sometimes wholly membranous. *Flowers* of various colours. *Perianth tube* short, rarely longer than the segments. *Capsules* shortly cylindrical or rarely subglobose, on straight or curved peduncles.

This section comprises 31 Cape species and probably also the species of the northern hemisphere. The Cape species have previously been placed in four subsections of the section *Romulea,* as well as in section *Pratenses* (De Vos, *l.c.*). These subsections and section *Pratenses* differ mainly in corm shape and chromosome number.

Widely distributed throughout the western, south-western and south-eastern Cape Province from Namaqualand to the Cape Peninsula and to Grahamstown and Bathurst.

1. **Romulea schlechteri** *Bég.* in Bot. Jb. 38: 335 (1907); in Malpighia 23: 93 (1909); De Vos in Jl S. Afr. Bot. Suppl. 9: 62, figs 8 & 13 (1972). Syntypes: Cape, Clanwilliam, Pakhuisberg, *Schlechter* 8648 (B, lecto.!; BOL!; GRA! BM!; G!; K!; Z!); near Hopefield, *Bachmann* 1576 (B!, partly).

R. × *hybrida* Bég. in Bot. Jb. 38: 339 (1907); in Malpighia 23: 107 (1909). Type: Cape, Caledon, Swartberg, *Zeyher* 4043 (G, holo.!; K!; P!).

R. elegans Klatt var. *parviflora* Bak. in F.C. 6: 42 (1896). Type: Cape, Zwartberg near Caledon, *Zeyher* 4043 (K, holo.!; G!; P!).

Plants 80− 450 mm long. *Corm* with a crescent-shaped basal ridge. *Stem* 20−300 mm long, hidden or extended above-ground. *Basal leaves* 2 or more, filiform or compressed cylindrical, 80−450 × 0,5−1,5 mm, grooves narrow, rib margins some-times minutely ciliate. *Bracts* green, inner with colourless membranous margins. *Flo-wers* 20−50 mm long, pale violet, lilac-pink or cream, cup golden or orange-yellow, outer perianth segments purplish, greenish or irregularly blotched or striped on the backs. *Perianth tube* 3−7 mm long; seg-ments 12−40 × 4−14 mm. *Filaments* 4−10 mm, orange-yellow; anthers 4−9 mm long, pale yellow. *Style* 9−15 (−20) mm; stigmas below to above the anther tips. *Capsules* on erect or suberect peduncles. *Chromosome no.* 2n=24.

Found in western Cape districts from Vanrhyns-dorp to Malmesbury and Worcester, and to Caledon. Map 2.

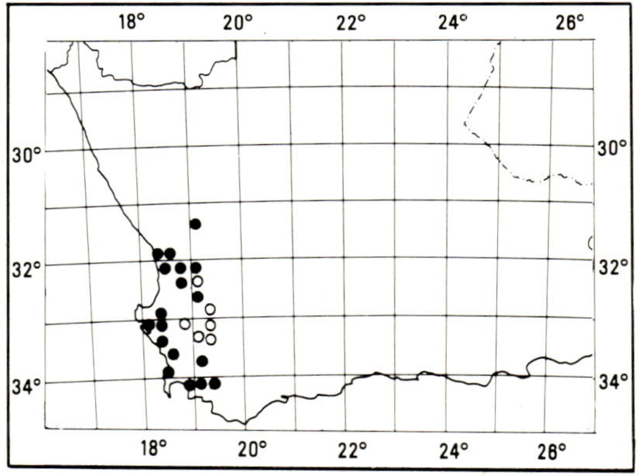

MAP 2.—● **Romulea schlechteri**
  ○ **R. saxatilis**

Vouchers: *Leipoldt* 3828 (BOL); *Gillett* 3671 (BOL); *De Vos* 1278 (STE); *Sidey* (SAM 65019); *Acocks* 23984.

A variable species distinguished by its corm, two basal leaves in long-stemmed forms, green bracts, the inner with wide white membranous margins, and variously coloured flowers. Three ecological races occur: (1) A form with pink or pale violet flowers and short stems on mountain slopes in the northern districts; (2) with cream flowers and tall stems on sandy plains from Hopefield to Malmesbury; (3) with cream or pinkish flowers on rather small plants on the Worcester to Caledon mountains.

2. **Romulea papyracea** *Wolley-Dod* in J. Bot., Lond. 38: 170 (1900); Bég. in Malpighia 23: 94 (1909); G. J. Lewis in Adamson & Salter, Fl. Cape Penins. 223

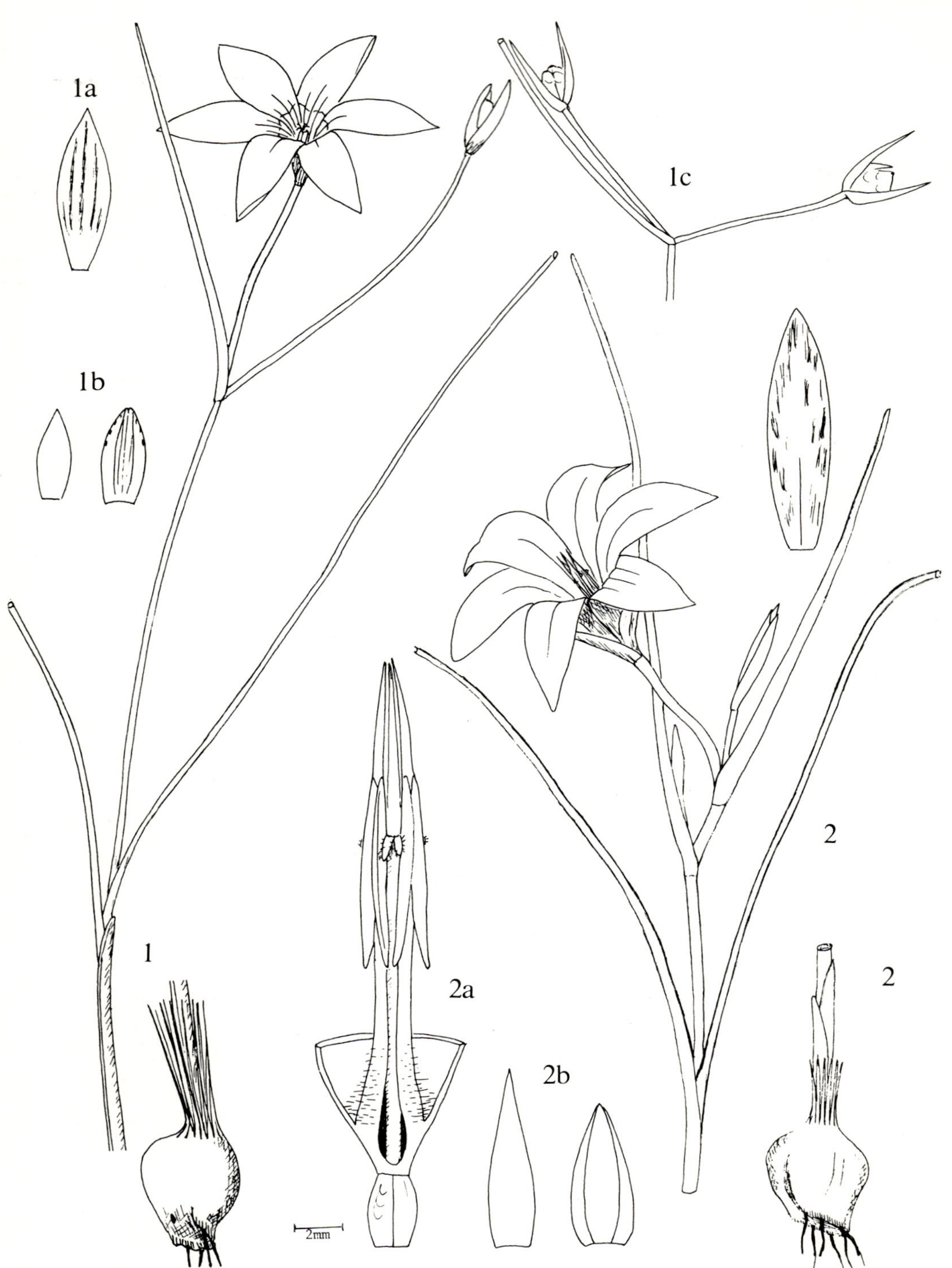
1a
1b
1c
1
2a
2b
2
2
2mm

(1950); De Vos in Jl S. Afr. Bot. Suppl. 9: 66 (1972). Type: Cape, Cape Peninsula, Table Mountain lower plateau, *Wolley-Dod* 3075 (BOL, holo.!; BM!; K!).

Closely related to *R. schlechteri* (no. 1) from which it differs as follows: *Stem* very short or up to 40 mm long. *Leaves* basal or basal and cauline, compressed cylindrical, recurved, up to 2 mm in diam., with slightly wider grooves and wider lateral ribs. *Outer bract* keeled, with prominent closely spaced veins, acuminate; inner shorter than the outer. *Flowers* only c. 25 mm long, pale lilac-pink, with a yellow cup, outer perianth segments darker pink on the backs. *Perianth tube* 4−5 mm long; segments c. 16 × 5 mm. *Filaments* shorter than the anthers which are 5 mm long. *Style* less than 10 mm, with stigmas not reaching the anther tips.

Found only once in 1897 on the lower plateau of Table Mountain near Cape Town (3318−CD).

Voucher: *Wolley-Dod* 3075 (BOL; BM; K).

The name is derived from the thin papery corm tunics, the hard outer tunical layers having been lost. One specimen in K has a corm with an indication of a crescent-shaped basal ridge.

3. **Romulea saxatilis** De Vos in Jl S. Afr. Bot. Suppl. 9: 71, fig. 16 (1972). Type: Cape, Ceres, lower mountain slopes, *Guthrie* 18558 (BOL, holo.!).

Very closely related to *R. schlechteri* (no. 1) differing in the following: *Plants* sometimes up to 600 mm long. *Stem* extended above ground, very slender, often somewhat bent or flexuose, 100−250 mm long. *Basal leaves* usually slender, flaccid, less than 1 mm in diam., glabrous. *Inner bracts* sometimes brown-dotted on the largely colourless membranous margins. *Flowers* 15−30 mm long, magenta-pink, sometimes with small dark blotches in the throat, cup golden-yellow. *Perianth segments* elliptical, 9−22 × 4−6 mm. *Filaments* 3−4 mm, minutely pilose almost to the tops, widened in the middle, as long as the anthers. *Style* 7−10 mm; stigmas at or above the anther tips. *Capsules* globose or ellipsoid, on straight spreading peduncles. Fig. 4:1.

Found in western Cape districts from Clanwilliam to Piketberg and Ceres, mainly on mountain slopes and plateaux amongst rocks. Map 2.

Vouchers: *Esterhuysen* 12171 (BOL; K); *Esterhuysen* 16162 (BOL); *Lewis* 2572 (SAM); *De Vos* 2055, 2053 (STE); *Pillans* 7715 (BOL).

This species differs from *R. schlechteri* (no. 1) mainly in its more slender habit, mostly elongated stem, flowers with shorter stamens and filaments widened in their middle and pilose almost to the tips, as well as in chromosome number (2n=c. 28).

4. **Romulea flexuosa** Klatt in Abh. Naturforsch. Ges. Halle 15: 400 (1882); Bak. Handb. Irid. 104 (1892); in F.C. 6: 42 (1896) pro syn.; Bég. in Malpighia 23: 117 (1909); De Vos in Jl S. Afr. Bot. Suppl. 9:66, fig. 14 (1972). Type: Cape, *Drège* 4038 partly, spec. 1 & 4 (S, holo.!).

R. attenuata De Vos in Jl. S. Afr. Bot. 21: 102 (1955); Type: Cape, Clanwilliam, Elandsfontein, *Leipoldt* 4247 (BOL, holo.!; PRE!).

Plants 150−400 mm long. *Corm* with a crescent-shaped basal ridge. *Stem* 20−250 mm long, usually extended above-ground, sometimes flexuose. *Basal leaves* 2, filiform, 150−400 × 0,5−1 mm, grooves very narrow. *Inner bract* green, with colourless membranous margins. *Flowers* 30−45 mm long, white, cup brownish or off-white, outer perianth segments on the backs with irregular pink or purplish brown and green markings. *Perianth tube* 5−8 mm long; segments elliptical (outer) and oblanceolate (inner), 25−35 × 7−11 mm. *Filaments* 6−7 mm; anthers about twice as long as the filaments, with attenuate connectives elongated 2,5−6 mm above the thecae. *Style* 12−16 mm; stigmas about halfway up the anthers. *Capsules* ellipsoidal on arcuate peduncles. *Chromosome no.* 2n=24. Fig. 4:2.

Found in western Cape coastal districts from Vanrhynsdorp to Piketberg, and also in the Hottentots-Holland region, on mountain plateaux amongst rocks and in crevices. Map 3.

Vouchers: *Acocks* 18224; *Oliver* STE 30273 (STE); *Salter* 7291 (BOL); *Leipoldt* 4247 (BOL; PRE); *Stokoe* 4576 (BOL).

An early flowering species (May to July), differing from other species in its elongated connectives and therefore longer anthers.

FIG. 4.−1, **Romulea saxatilis,** habit, × 1; 1a, outer perianth segment, lower face; 1b, outer (left) and inner (right) bract; 1c, mature capsules (*De Vos* 2053). 2, **R. flexuosa,** habit, × 1; 2a, perianth tube, stamens and pistil; 2b, outer (left) and inner (right) bract (STE 30273).

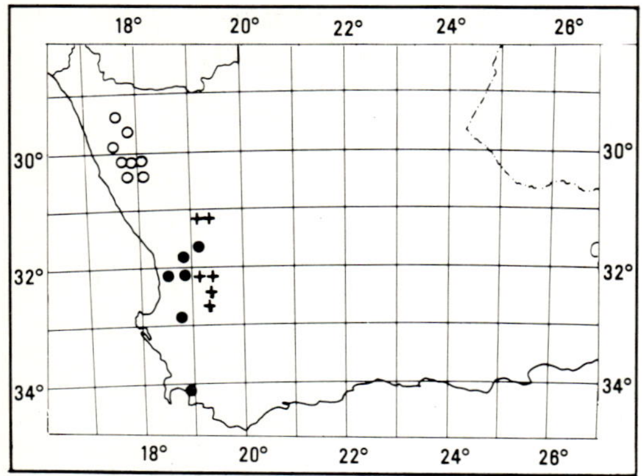

MAP 3.—● **Romulea flexuosa**
　　　○ R. namaquensis
　　　+ R. montana

5. **Romulea vinacea** *De Vos* in Jl S. Afr. Bot. Suppl. 9: 69, fig. 15 (1972). Type: Cape, Clanwilliam, top of Pakhuis Pass, *Lewis* 2120 (SAM, holo.!; PRE, iso.!).

Plants 70−240 mm long. *Corm* with a crescent-shaped basal ridge. *Stem* 20−70 mm, usually shortly extended above-ground. *Basal leaves* 2, filiform, 60−240 × up to 1 mm, often minutely ciliate on the rib margins, grooves narrow. *Bracts* largely green, inner with colourless membranous margins which are brown-edged in the upper half. *Flowers* 20−40 mm long, light bluish-violet with darker violet veins, cup cream with yellow markings, outer segments wine-coloured on the backs. *Perianth tube* 4−5 (−8) mm long; segments 14−28 × 6−10 mm, mostly obtuse. *Filaments* 6−9 mm, pilose almost to their tips; anthers 4−6 mm long. *Style* 10−14 mm; stigmas at the anther tips. *Capsules* subglobose to ellipsoid on suberect peduncles. *Chromosome no.* 2n=24.

Found only in the Pakhuis Pass, Clanwilliam, near the summit and around Leipoldt's grave, in sandy spots (3219−AA).

Vouchers: *Leipoldt* in BOL 21278; *Barker* 6591 (NBG); *Lewis* 2010 (SAM); *De Vos* 1921, 2108 (STE).

A rare species with almost lavender coloured flowers which open only around 15h00. The outer segments are shiny, and plum- or wine- coloured on the backs.

6. **Romulea biflora** *(Bég.) De Vos* in Jl S. Afr. Bot. Suppl. 9: 75, figs 7 & 17 (1972).

R. ambigua Bég. var. *biflora* Bég. in Malpighia 23: 80 (1909). Type: Cape, Bidouwberg, Clanwilliam, *Schlechter* 8694 (G, holo.!; GRA!; PRE!; B!; BM!; K!; S!).

Plants 100−250 (−300) mm long. *Corm* with an almost circular basal ridge. *Stem* 20−150 mm long. *Basal leaves* usually 2, filiform, 70−300 × 0,5−1,5 mm, grooves usually narrow. *Inner bract* green or greenish, with brown-streaked or sometimes colourless membranous margins. *Flowers* 25−45 mm long, deep old rose or bright pink with large purple or violet blotches in the throat and a small dark spot on each side of the segments, cup golden-yellow, outer segments striped or mottled on the backs. *Perianth tube* 4−5 mm long; segments 18−35 × 6−10 mm. *Filaments* 5−7 mm; anthers as long, yellow or rarely violet. *Style* 11−13 mm; stigmas pale or purple, at the anther tips or higher. *Capsules* subglobose or ellipsoid, on suberect or bent peduncles. *Chromosome no.* 2n=24.

Found in the western Cape districts of Vanrhyns-dorp and Clanwilliam, on red clayey higher ground (3118−DB, DC; 3219−AA, AB).

Vouchers: *Leipoldt* in BOL 20770 (BOL; SAM); *Acocks* 19296 PRE; K; M); *Marsh* 394 (STE); *Oliver* 4967 (STE); *De Vos* 2111 (STE).

Distinguished from the five above-mentioned species by its flowers with large purple or violet blotches in the throat, inner bract usually with brown-streaked membranous margins (sometimes co-lourless), and corms with an almost circular basal ridge.

7. **Romulea namaquensis** *De Vos* in Jl S. Afr. Bot. 21: 103, fig. 2 (1955); idem, Suppl. 9: 76 (1972). Type: Cape, Ka-miesberg between Witsand and Leliefon-tein, *Pearson sub P. Sladen Mem. Exp.* 6656 (BOL, holo.!; K!).

R. namaquensis subsp. *bolusii* De Vos, idem, Suppl. 9: 77, fig. 18 (1972). Type: Cape, near Okiep, *Bolus* 6620 (STE, holo.!; BOL!; K!).

FIG 5.−1, **Romulea namaquensis**, habit, × 1; 1a, perianth tube, stamens and pistil; 1b, outer (left) and inner (right) bract (*De Vos* 2173). 2, **R. kamisensis**, habit, × c.1; 2a, outer perianth segments: upper face (left) and lower face (right); 2b, outer (left) and inner (right) bracts; 2c, perianth tube, stamens and pistil; 2d, ripening capsules; 2e, transverse section of leaf (*De Vos* 2232).

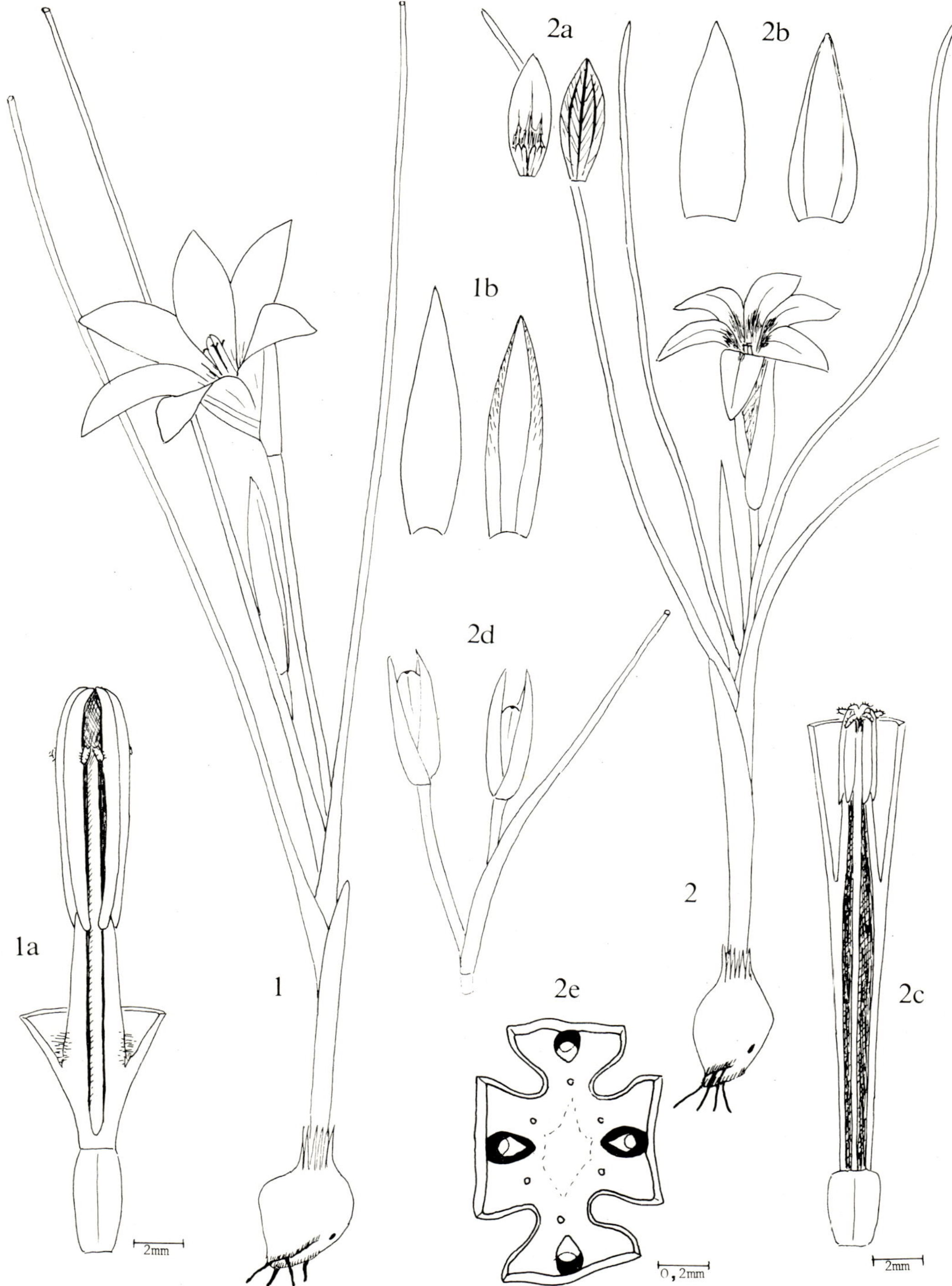
2a
2b
1b
2d
2
1a
1
2e
2c
2mm
0,2mm
2mm

Plants 70−200 mm long. *Corm* with a crescent-shaped basal ridge. *Stem* short, hidden, or up to 80 mm long. *Basal leaves* 2 or more, filiform 70−200 × 0,5−1 mm, grooves very narrow. *Inner bract* green with colourless or brown-edged or streaked membranous margins. *Flowers* 20−50 mm long, shiny, rose to salmon-pink, sometimes almost white, with small blotches or dark veins in the throat, outer perianth segments irregularly blotched on the backs. *Perianth tube* 4−12 mm long; segments 16−40 × 5−10 mm. *Filaments* 4−10 mm, yellow or maroon, inserted above the middle of the perianth tube, subequal to the golden yellow anthers. *Style* 8−20 mm; stigmas lower or higher than the anther tips. *Capsules* shortly cylindrical, on suberect peduncles. *Chromosome no.* 2n=26. Fig. 5:1.

Widespread in Namaqualand from Okiep to Kamieskroon and the Kamiesberg. Map 3.

Vouchers: *Esterhuysen* 5453 (BOL); *Marloth* 13272 (PRE; STE); *Bolus* 6620 (BOL; STE; K); *De Vos* 1617 (STE); *Compton* 11170 (NBG).

Distinguished by its usually pink flowers with very little yellow in the base, and with maroon filaments (rarely yellow) inserted just above the middle of the rather long perianth tube. Variation occurs in the length of the flower and its parts, especially the perianth tube. Previously two subspecies were established, based mainly on differences in the size of the floral organs. The variation, however, is continuous, and subsp. *bolusii* has now been merged with the typical subspecies.

8. **Romulea kamisensis** *De Vos* in Jl S. Afr. Bot. Suppl. 9: 81, figs 9 & 19 (1972). Type: Cape, Studer's Pass (Platbakkies and Garies), *Stayner* s.n., 26-8-1968 (NBG, holo.!).

Plants 100−150 mm long. *Corm* with a small crescent-shaped basal ridge. *Stem* short or up to 50 mm long. *Basal leaves* 2 or more, filiform, 100−150 × 0,5−1 mm, grooves very narrow. *Outer bract* green, up to 22 mm long, reaching to the bases of the perianth segments; inner green with colourless membranous margins. *Flowers* 30−40 mm long, magenta-purple, with a violet blotch and violet stripes on each perianth segment in the throat. *Perianth tube* 17−22 mm long, long-funnel-shaped; segments 11−16 × 3,5−5 mm. *Stamens* included; filaments inserted above the middle of the perianth tube, 4−5 mm long,

as long as the golden-yellow anthers. *Style* 15−16 mm; stigmas more or less at the anther tips. *Capsules* cylindrical, on suberect peduncles. *Chromosome no.* 2n=c.26. Fig. 5:2.

Found in Namaqualand on the Kamiesberg and around Kamieskroon and Garies, on sandy loam (3018−AA, AC, CA/CB).

Vouchers: *Leipoldt* 3832 (BOL); *Pearson* 6671 (BOL); *De Vos* 2232 (STE; PRE).

Readily distinguished by its long perianth tube which is longer than the segments, with stamens and style included in the tube.

9. **Romulea elliptica** *De Vos* in Jl S. Afr. Bot. Suppl. 9: 83, fig. 20 (1972). Type: Cape, between Vredenburg and Saldanha, *De Vos* 2226 (STE, holo.!; PRE).

Plants 150−300 mm long. *Corm* with a small crescent-shaped basal ridge. *Stem* short or up to 160 mm long. *Basal leaves* 2, filiform, 150−300 × 1−1,5 mm, suberect, grooves narrow. *Inner bract* largely green, with colourless membranous margins and submembranous base. *Flowers* 25−35 mm long, bright golden-yellow, with some dark veins in the cup. *Perianth tube* 4−5 mm long; segments elliptical, mostly obtuse, 18−27 × 5−9 mm, outer segments slightly longer than the inner, green on the backs, with brown markings along the margins. *Filaments* 6−7 mm, minutely pilose almost to the tops; anthers 4−6 mm long, golden-yellow. *Style* 10−12 mm; stigmas at or above the anther tips. *Capsules* cylindrical, on suberect peduncles. *Chromosome no.* 2n=24.

Found in a single locality between Vredenburg and Saldanha, on white sandy flats (3217−DD).

Vouchers: *De Vos* 2017, 2226 (PRE; STE).

Related to the white-flowered form of *R. schlechteri* (no. 1) and to the yellow *R. saldanhensis* (no. 16), but differing in the shape of the perianth segments and the more pilose filaments; and from the latter also in its two basal foliage leaves and greener inner bract.

10. **Romulea sulphurea** *Bég.* in Bot. Jb. 38: 331 (1907); in Malpighia 23: 100 (1909); De Vos in Jl S. Afr. Bot. Suppl. 9: 85 (1972). Type: Cape, Pakhuis Pass, Clanwilliam, *Schlechter* 10818 (Z, lecto.!; BOL!; GRA!; PRE!; B!; BM!; K!; S!; etc.).

Plants 50−150 mm long. *Corm* with a small crescent-shaped basal ridge. *Stem* up to 35 mm long, hardly extended from the leaf bases. *Basal leaves* 2 or more, filiform, 50—150 mm long, less than 1 mm wide, with ribs and grooves hardly visible. *Inner bract* green, with brown-speckled membranous margins. *Flowers* 15—25 mm long, sulphur-yellow, often with dark linear-oblong marks in the cup, outer segments with a purple spot on each side at the base. *Perianth tube* 4—5 mm long; segments 12—20 × 4—5 mm, subacute to subobtuse. *Filaments* 6—9 mm; anthers 2,5—3 mm, yellow. *Style* 8—11 mm; stigmas reaching the anther tips. *Capsules* obovoid.

Found only once, in 1897, in the Pakhuis Pass above Clanwilliam, amongst rocks (3219—AA).

Vouchers: *Schlechter* 10818 (BOL; PRE; GRA; K; BM; Z; etc.).

Labelled *R. aurea* Schltr. (ined., non Klatt), in some herbaria, and *R. sublutea* Bak. var. *sulphurea* Bég. (ined.) in B.

Distinguished by its very short anthers which reach halfway or higher up the perianth. Of the dark marks recorded by Béguinot on the perianth, only the lateral markings at the bases of the segments are now visible in herbarium specimens.

11. **Romulea montana** *Schltr. ex Bég.* in Bot. Jb. 38: 332 (1907); in Malpighia 23: 100 (1909); De Vos in Jl S. Afr. Bot. Suppl. 9: 87, fig. 21 (1972). Type: Cape, Onder-Bokkeveld, Oorlogskloof, *Schlechter* 10949 (G, holo.!; BOL!; GRA!; PRE!; B!; K!; S!; etc.).

R. hirsuta (Eckl. ex Klatt) Bak. var. *aurantiaca* Schltr. in Bot. Jb. 27: 90 (1900). *R. ambigua* Bég. var. *aurantiaca* (Schltr.) Bég. in Malpighia 23: 80 (1909). Type: Cape, Matjiesrivier, Clanwilliam, *Schlechter* 8847 (B, holo.!; BOL!; GRA!; PRE, partly!; BM!; K!; Z!; etc.).

R. rosea Eckl. var. *flavescens* Bég. in Malpighia 23: 63 (1909). Type: Cape, without locality, *Mundt & Maire* 580 (B, holo.!).

Plants 80—300 mm long. *Corm* with a wide crescent-shaped basal ridge sometimes forming an almost complete circle. *Stem* up to 150 mm long or short and hidden. *Basal leaves* 2 or more, filiform, 60—300 × 0,5—1 mm, grooves rather narrow. *Bracts* green, inner with brown-edged or brownish membranous margins. *Flowers* 20—45 mm long, shiny buttercup-yellow, with a dark brown blotch sometimes reduced to dark veins on each segment in the throat, outer segments reddish brown on the backs or with faint feathered veining. *Perianth tube* 4—6 mm

long; segments 15—35 × 5—10 mm, obtuse to subacute. *Filaments* 5—6 mm; anthers 4—8 mm long, yellow. *Style* 10—15 mm; stigmas about at the anther tips. *Capsules* shortly cylindrical, on widely patent peduncles. *Chromosome no.* 2n=24. Fig. 6:1.

Found on the western escarpment from Calvinia to Clanwilliam, in stony ground at altitudes of 600 m and more. Map 3.

Vouchers: *Schlechter* 8002 (BOL; GRA; PRE; K; BM); *Bolus* in BOL 20519 (BOL); *Lewis* 2013 (SAM); *Lorenzo* 22 (STE); *Barker* 6478 (NBG; STE).

Distinguished by its corm with a crescent-shaped basal ridge which is usually wider than the corm itself, and by bright yellow flowers, often with dark blotches in the throat.

12. **Romulea toximontana** *De Vos* in Jl S. Afr. Bot. Suppl. 9: 89, fig. 22 (1972). Type: Cape, plateau on Gifberg, Vanrhyns-dorp, *De Vos* 2020 (STE, holo.!; PRE).

Closely related to *R. montana* (no. 11) from which it differs as follows: *Stem* up to 100 mm long. *Basal leaves* sometimes minutely and sparsely ciliate on the rib margins. *Flowers* 18—30 mm long, cream with an orange cup, the outer segments on the backs green or purple or irregularly blotched. *Perianth segments* 13—22 mm long. *Filaments* 3—5 mm, orange-yellow; anthers pale yellow. *Style* 7—10 mm. *Capsules* ellipsoid, on widely patent straight peduncles.

Found on the plateau of the Gifberg near Vanrhynsdorp and on the escarpment above Van-rhyn's Pass, on sandy soil at c. 500—750 m altitude (3118—DC; 3119—AC).

Vouchers: *Lewis* 2005 (SAM); *Marsh* 510 (STE); *De Vos* 2176 (STE).

The shape of the corm is similar to that of *R. montana* (no. 11) but *R. toximontana* differs in its cream-coloured flowers with shorter style, in its chromosome number (2n=c. 28, compared with 24 in *R. montana*) and in peduncles diverging at an angle of about 90° in the fruiting stage. From the sympatric *R. sladenii* (no. 24) it differs in corm shape, chromosome number and in the smaller divergence of the fruiting peduncles.

13. **Romulea tabularis** *Bég.* in Bot. Jb. 38: 337 (1907); in Malpighia 23: 104 (1909); De Vos in Jl S. Afr. Bot. Suppl. 9: 93, figs 10 & 23 (1972). Type: Cape, C.B.S., *Ecklon & Zeyher* 595 sub Irid. 199 (B, holo.!).

R. bulbocodioides sensu Eckl., Top. Verz. 19 (1827), excl. syn., non Bak., nec Klatt.

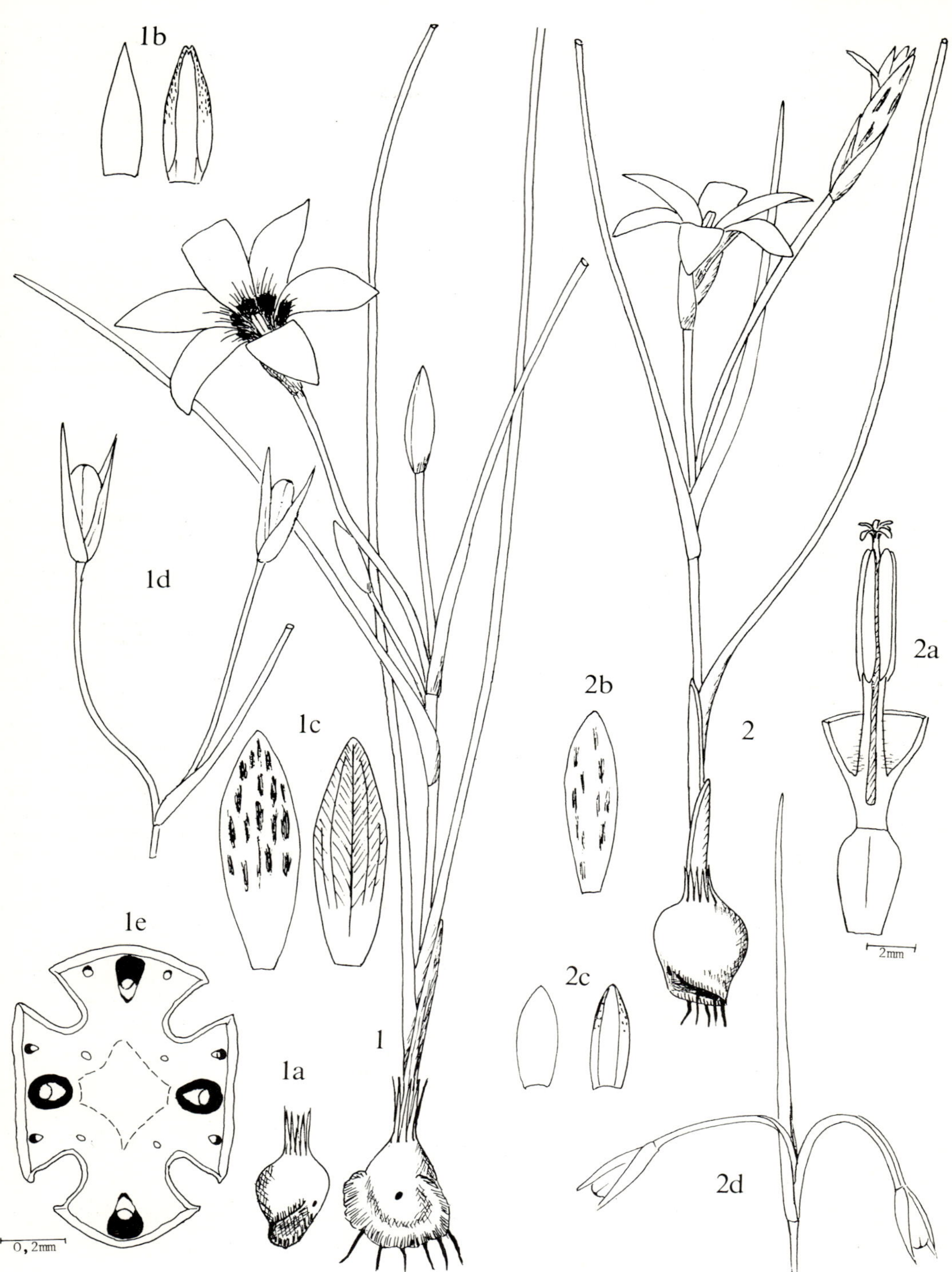

R. rosea var. *parviflora* Bak., Handb. Irid. 104 (1892), partly and in F.C. 6: 42 (1896), partly. Type: Cape, without locality, Herb. *Drège* sub *Trichonema recurvum b* (K partly, holo.!).

R. versicolor Bég. in Malpighia 23: 116 (1909), excl. syn. Type: Cape, Malmesbury, near Darling, *Schlechter* 5340 (BOL, lecto.!; GRA!; K!; Z!).

R. duthieae L. Bol. in S. Afr. Gdng 18: 342 (1928). Type: Cape, Stellenbosch flats, *Duthie* s.n. (BOL, holo.!; STE!).

Plants 100−350 (−600) mm long. *Corm* with a crescent-shaped basal ridge. *Stem* 10−100 (−300) mm long. *Basal leaves* 1,2 or more, filiform, or compressed cylindrical, 100−350 (−600) × 1−2 mm, grooves narrow. *Inner bract* greenish or reddish, with brown-spotted or brown-edged membranous margins. *Flowers* 15−35 mm long; lavender-blue or sometimes white or bluish-mauve, in the lower half orange or golden yellow, often with a pale transverse band in the throat, outer segments on the backs irregularly marked in green and brown. *Perianth tube* 3−5 mm long; segments narrowly elliptical, 10−28 × 3−7 mm, obtuse or subobtuse. *Filaments* 3−6 mm; anthers 3−6 mm, yellow. *Style* 8−13 mm; stigmas at or above the anther tips. *Capsules* shortly cylindrical, on curved peduncles which straighten later. *Chromosome no.* 2n=24. Fig. 6:2.

Widely distributed in western and south-western Cape districts from Clanwilliam to the Cape Peninsula and to Bredasdorp, generally at low altitude in moist, sandy or clayey areas. Map 4.

Vouchers: *Leipoldt* 3050 (BOL; PRE); *Salter* 6160 (BOL); *Esterhuysen* 18867 (BOL); *Lewis* 3567 (SAM); *De Vos* 2279 (STE).

Readily distinguished in the veld by its generally lavender-blue flowers with brown blotches on the backs of the outer perianth segments. This colouring usually fades in herbarium specimens, and such specimens, as well as the cream-coloured variants, can be distinguished from the closely related *R. leipoldtii* (no. 14) only by their slightly smaller flowers with shorter style and usually narrower segments, and by their slightly more membranous inner bracts. Colour variants occur at Kreeftebaai, Hopefield district (cream) and between Velddrif and Aurora, Piketberg district (blue-mauve).

14. **Romulea leipoldtii** *Marais* in Curtis's bot. Mag. 175: t. 460 (1964); De Vos in Jl S. Afr. Bot. Suppl. 9: 96 (1972). Type:

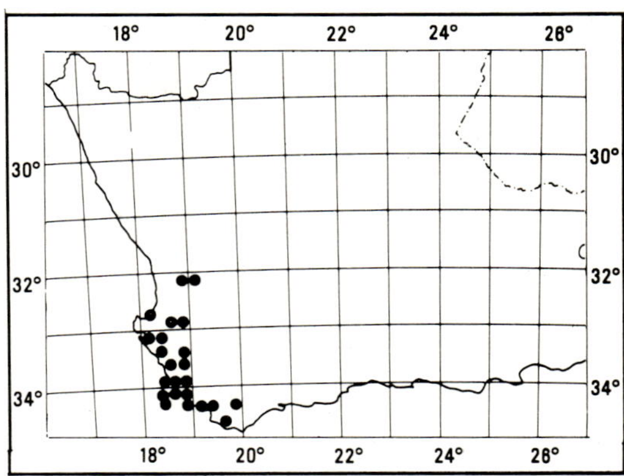

MAP 4.— **Romulea tabularis**

Cape, Warmbaths, *Leipoldt* s.n. (K, holo.!). The topotype, *Leipoldt* in BOL 20487, is probably an isotype (BOL!; SAM!).

Trichonema filifolium sensu Klatt in Linnaea 34: 671 (1865-66), excl. syn. Eckl., non Ker-Gawl. (1827).

Romulea filifolia sensu Bak. in J. Linn. Soc., Bot. 16: 88 (1877); Handb. Irid. 101 (1892); in F.C. 6: 38 (1896); sensu Klatt in Abh. Naturforsch. Ges. Halle 15: 402 (1882); sensu Bég. in Malpighia 23: 101 (1909), excl. syn. *R. tubata* and *R. schlechteriana*, non Eckl. (1827).

Very closely related to *R. tabularis* (no. 13) from which it differs as follows: *Stem* 50−350 mm long. *Basal leaves* usually 2, c. 1 mm in diam. *Inner bract* green or greenish with brown-spotted or -edged or almost colourless membranous margins. *Flowers* 25−40 mm long, cream or white in the upper half of the perianth segments with yellow veins, outer segments sometimes greenish or fawn on the backs. *Perianth tube* up to 8 mm long; segments 18−35 × 5−8 mm, subacute to obtuse. *Filaments* 5−8 mm; anthers 5−8 mm long. *Style* 12−20 mm. *Chromosome no.* 2n=24.

Found in the western Cape districts from Clanwilliam to Malmesbury, in damp sandy localities. Map 5.

Vouchers: *Leipoldt* 3541 (BOL; PRE); *Acocks* 19781 (PRE; K; M); *Barker* 4763 (NGB); *De Vos* 1466 (STE); *Taylor* 5937 (STE).

FIG. 6.−1, **Romulea montana,** habit, × c.1; 1a, younger corm; 1b, outer (left) and inner (right) bract; 1c, outer perianth segments of two plants, lower face; 1d, almost mature capsules; 1e, transverse section of leaf (*De Vos* 1925). 2, **R. tabularis,** habit, × 1; 2a, perianth tube, stamens and pistil; 2b, outer perianth segment, lower face; 2c, outer (left) and inner (right) bracts; 2d, ripening capsules (*De Vos* 1690).

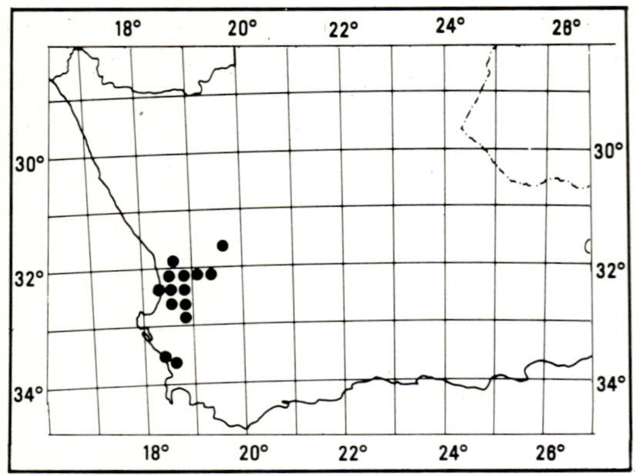

MAP 5.— **Romulea leipoldtii**

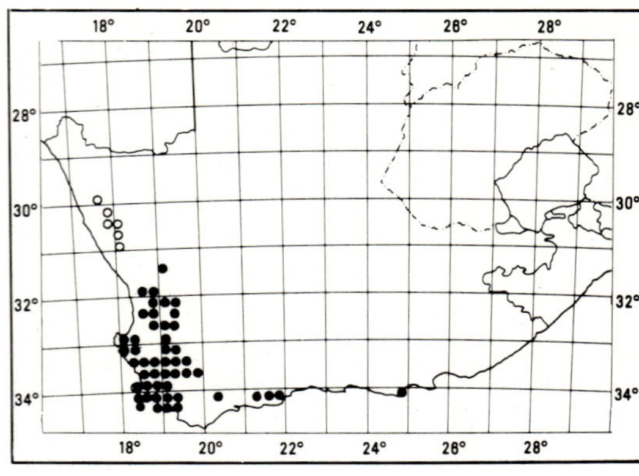

MAP 6.—● **Romulea flava**
○ **R. citrina**

The main distinguishing features between this species and faded herbarium specimens of *R. tabularis* (no. 13) are the slightly larger flowers and less membranous bracts of the present species.

15. **Romulea flava** *(Lam.) De Vos* in Jl S. Afr. Bot. 36: 273 (1970), ibid. Suppl. 9: 98 (1972); Goldbl. & Barnard in Jl S. Afr. Bot. 36: 306 (1970). Type: Without locality, without collector in Herb. Lamarck (P, lecto.!).

Plants 100−550 mm long. *Corm* with a crescent-shaped basal ridge. *Stem* short, hidden, or up to 300 mm long. *Basal leaf* single in long-stemmed forms, filiform or compressed cylindrical, 100−550 × 0,8−3(−4) mm, suberect or curved, sometimes minutely ciliate on the rib margins, grooves wide or narrow. *Inner bract* membranous or with the centre sometimes submembranous. *Flowers* (18−)20−40 mm long, yellow or white, sometimes blue, blue-violet or rarely pink, cup yellow or greenish yellow, outer segments green or greenish brown on the backs. *Perianth tube* 3−7 mm long; segments 10−30 × 3−12 mm. *Filaments* 4−7 mm; anthers 4−7 mm, golden yellow. *Style* 8−15 mm; stigmas more or less at the anther tips. *Capsules* shortly cylindrical, on recurved peduncles which straighten later. *Chromosome no.* 2n =24(48).

Widely distributed from Vanrhynsdorp to the Cape Peninsula and to Riversdale, with outliers to Namaqualand and to Humansdorp. Map 6.

A very common, polymorphic species readily distinguished by its green outer bract with closely spaced veins, membranous or sometimes submembranous inner bract and a single basal leaf in long-stemmed forms. Four varieties are recognized:

1 Leaves glabrous:

  2 Flowers 25 mm long or longer; yellow to white, sometimes blue .... (a) var. *flava*

  2 Flowers less than 25 mm long, yellow .... (b) var. *minor*

1 Leaves usually minutely ciliolate on the rib margins:

  3 Stem short; leaves 2−4 mm in diameter .... (d) var. *hirsuta*

  3 Stem often elongated; leaves c. 1 mm in diameter .... (c) var. *viridiflora*

(a) var. **flava.**

Ixia flava Lam., Tab. Encycl. 109 (1791), non Hornem. (1819). *Bulbocodium flavum* (Lam.) Kuntze, Rev. Gen. 2: 700 (1891).

Ixia bulbocodium sensu Thunb., Diss. Ixia 6 (1783), partly; Fl. Cap. 1: 218 (1811), non Murray, as to: Cape, C.B.S., *Thunberg* (UPS).

Trichonema pudicum Ker-Gawl. in König & Sims, Ann. Bot. 1: 223 (1805), nom. nud., non Ker-Gawl. (1810), nec. Steud. (1841), nec Klatt (1865−66). Based on *Solander* s.n. (BM!).

Ixia recurva F. Delaroche in Red., Liliac. 5: t. 251, fig. 1 (1809). *Trichonema recurvum* (F. Delaroche) Spreng., Syst. Veg. 1: 149 (1825); Ker-Gawl., Irid. Gen. 83 (1827). *R. recurva* (F. Delaroche) Eckl., Top. Verz. 20 (1827). *Ixia recurvifolia* Poir. in Lam., Encycl. Suppl. 3: 201 (1813). *Trichonema recurvifolium* (Poir.) Ker-Gawl., Irid. Gen. 83 (1827). *Geissorhiza recurvifolia* (Poir.) Klatt in Linnaea 34: 655 (1865−66), partly, as to name, but excl. spec. cit. and syn. Eckl. & Vahl. Iconotype: Red. Liliac. 5: t. 251 fig. 1 (1809) (lecto.!).

Ixia reflexa Thunb., Fl. Cap. 1: 220 (1811), non Andr. (1797). *Trichonema reflexum* (Thunb.) Steud., Nom. Bot. edn 2, 2: 702 (1841), excl. syn. Eckl. Type: Cape, C.B.S., *Thunberg* s.n. herb. no. 984 (UPS, lecto.!).

R. chloroleuca (Jacq.) Eckl., Top. Verz. 20 (1827), excl. syn. Jacq., non Bak. (1877), nec Klatt (1882). Type: Cape, C.B.S. *Ecklon* 202 (S, holo.!).

R. fragrans Eckl., Top. Verz. 19 (1827), excl. syn. Type: Cape, C.B.S., Tafelberg, *Ecklon* 704 (S, holo.!; PRE!).

?R. candida Ten., Cat. Ort. Nap. 95 (1845). No type specimen found.

R. latifolia Herb. ex Bak. in J. Bot. Lond. 5: 237 (1876); in F.C. 6: 39 (1896). *Bulbocodium latifolium* (Bak.) Kuntze, Rev. Gen. 700 (1891). *R. bulbocodioides* var. *latifolia* (Bak.) Bég. in Annu. Conserv. Jard. bot. Genève 11–12: 163 (1908); in Malpighia 23: 109 (1909). Iconotype: *Herbert* MS. in Lindley Library, R. Hort. Soc., London, sub *Trichonema latifolium* W. Herb. ined.

R. bulbocodioides sensu Bak. in J. Linn. Soc., Bot. 16: 88 (1877); in F.C. 6: 37 (1896), excl. syn. Delaroche, Salisb. and Ker-Gawl.; Klatt in Abh. Naturforsch. Ges. Halle 15: 402 (1882), excl. same syn.; Bég. in Malpighia 23: 108 (1909), excl. same syn.; G.J. Lewis in Adams. & Salter, Fl. Cape Penins. 221 (1950), non Eckl. (1827). Type: Cape, near Cape Town, *Bolus* 2810 (K, lecto.!; BOL!; SAM!).

R. bulbocodioides var. *elongata* Thunb. ex. Bak. in F. C. 6: 38 (1896); Bég. in Malpighia 23: 110 (1909). *Ixia reflexa* Thunb., Fl. Cap. 1: 220 (1811) var. *elongata* Thunb., nom. nud. Type: *Thunberg* s.n. herb. no. 988 (UPS, holo.!).

R. bachmannii Bég. in Ann. Conserv. Jard. bot. Genève 11–12: 161 (1908); in Malpighia 23: 103 (1909), excl. var. Type: Cape, Clanwilliam, Modderfontein, *Schlechter* 7970 (G, holo.!; BOL!; GRA!; PRE!; BM!; Z!; etc.).

Stem short or up to 300 mm long. *Basal leaf* glabrous, 1–3(–4) mm wide, often with wide grooves. *Bracts* 12 mm or longer, the inner membranous. *Flowers* 25–40 mm, sulphur-yellow, cream or white, rarely blue. *Perianth tube* 4–7 mm long; segments 15–30 mm long. *Style* 10–15 mm.

From Clanwilliam to the Cape Peninsula and to Riversdale.

Vouchers: *Wolley-Dod* 2759 (BOL; BM; K); *Salter* 2623 (BOL; K); *Marloth* 9083 (PRE); *Esterhuysen* 18711 (BOL; PRE; NGB).

(b) var. **minor** *(Bég.) De Vos* in Jl S. Afr. Bot. Suppl. 9: 103 (1972). Syntypes: Cape, Clanwilliam, Koudeberg, *Schlechter* 8782 (G, lecto.!; PRE!; GRA!; BM!; K!; P!; G!); Clanwilliam, Langekloof, *Schlechter* 8396 (G, syn.!; PRE!).

R. bulbocodioides var. *minor* Bég. in Annu. Conserv. Jard. bot. Genève 11–12: 163 (1908); in Malpighia 23: 110 (1909).

Trichonema caulescens Ker-Gawl. in Curtis's bot. Mag. 34: t. 1392 (1811); Irid. Gen. 82 (1827), excl. syn. Delaroche & Lam. *R. caulescens* (Ker-Gawl.) Klatt in Abh. Naturforsch. Ges. Halle 15: 399 (1882), excl. syn. Lam. *T. hypoxidiflorum* Salisb. in Trans. hort. Soc. Lond. 1: 316 (1812). Iconotype: Curtis's bot. Mag. 34: t. 1392 (1811).

Differs from var. *flava* in the following: *Bracts* 8–12 mm long. *Flowers* yellow, less than 25 mm long, with shorter perianth tube (3–4 mm), segments (10–15 mm) and style (8–9 mm).

From Vanrhynsdorp to the Cape and to Humansdorp.

Vouchers: *Compton* 17361 (BOL); *Salter* 6822 (BOL; SAM); *Esterhuysen* 10287 (BOL; PRE).

(c) var. **viridiflora** *(Bég.) De Vos* in Jl S. Afr. Bot. Suppl. 9: 104 (1972). Syntypes: Cape, near Hopefield, *Bachmann* 1856 (B, lecto.!; Z!); same locality, *Bachmann* 1845 (B!, Z!).

R. bulbocodioides var. *viridiflora* Bég. in Malpighia 23: 110 (1909).

Ixia bulbocodium sensu Thunb., Diss. Ixia 6 (1783); Fl. Cap. 1: 218 (1811), partly, non Murray. Type: Cape, C.B.S., *Thunberg* (UPS).

Trichonema arenarium Eckl. ex Klatt in Linnaea 34: 667 (1865–66), excl. syn. *R. ramosa*. *R. arenaria* Eckl., Top. Verz. 18 (1827), nom. nud.; Bak. in J. Linn. Soc., Bot. 16: 89 (1877); in F.C. 6: 43 (1896), excl. syn. *R. ramosa*. *Bulbocodium arenarium* Kuntze, Rev. Gen. 2: 700 (1891). Type: Cape, Doornhoogte, Cape Flats, *Ecklon & Zeyher* Irid. 200 (B, lecto.!; K!, G!, PRE!).

R. similis Eckl. ex Bak., Handb. Irid. 102 (1892); in F.C. 6: 40 (1896), excl. syn.; Eckl., Top. Verz. 19 (1827), nom. nud. *Trichonema simile* Steud., Nom. Bot. edn 2, 2: 702 (1841), nom. nud. *Bulbocodium simile* (Eckl.) Kuntze, Rev. Gen. 2: 700 (1891). Syntypes: Cape, C.B.S., *Thunberg* (UPS!); *Oldenburg* 453 (BM!).

Differs from var. *flava* in the following: *Stem* frequently elongated. *Basal leaf* usually suberect, minutely ciliate on the rib margins, c. 1 mm diam., grooves narrow. *Inner bract* sometimes submembranous and greenish in the middle, with wide brown or brown-streaked membranous margins. *Flowers* with blue or white perianth segments.

From Clanwilliam to Caledon and in Namaqualand.

Vouchers: *Ecklon & Zeyher* Irid. 200 (B; G; K; PRE 22351); *Wolley-Dod* 1256 (BOL; K); *Boucher* 73 (STE); *Thompson* 98 (STE).

(d) var. **hirsuta** *(Bég.) De Vos* in Jl S. Afr. Bot. Suppl. 9: 105 (1972). Type: Cape, without locality, *Verreaux* s.n. anno 1831 (G, holo.!).

R. cruciata var. *hirsuta* Bég. in Annu. Conserv. Jard. bot. Genève 11—12: 158 (1908); in Malpighia 23: 68 (1909).

R. cruciata sensu Lewis in Adams. & Salter, Fl. Cape Penins. 222 (1950), non Eckl. (1827), nec. Bak. (1877).

R. vulgaris Eckl., Top. Verz. 18 (1827), excl. syn., nom. nud.

Differs from var. *flava* as follows: *Stem* short, hidden by the leaf bases. *Leaves* all basal, 2—4 mm wide, X-shaped in transverse section, minutely ciliate on the rib margins, usually recurved. *Inner bract* submembranous, sometimes greenish in the middle with brown membranous margins. *Flowers* with blue or violet-blue, or sometimes pink or white segments.

From Clanwilliam to Caledon.

Vouchers: *Salter* 8205 (BOL); *Bolus* 3734 (BOL; K, partly); *Wolley-Dod* 2659 (BOL; BM; K); *Johnson* 156 (NGB).

16. **Romulea saldanhensis** *De Vos* in Jl S. Afr. Bot. Suppl. 9: 108, fig. 26 (1972). Type: Cape, Saldanha, SW of Military Academy, *De Vos* 1772 (STE, holo.!).

Closely related to *R. flava* (no. 15) from which it differs as follows: *Plants* 200—600 mm long. *Basal leaf* filiform, 200—600 × 1—2 mm, grooves rather narrow. *Inner bract* submembranous, greenish in the middle of the upper half, with wide colourless or brown-streaked or speckled membranous margins. *Flowers* (20—)30—40 mm long, bright golden or cadmium-yellow, with slender dark lines in the cup, outer perianth segments frequently with irregular brown markings on the backs. *Perianth tube* slightly shorter and wider. *Style* 12—18 mm; stigmas at or above the anther tips. *Chromosome no.* 2n=24.

Found in the south-western Cape coastal districts of Vredenburg, Hopefield and Malmesbury, at low altitude on moist sandy or clayey ground (3217—DD; 3218—CC; 3317—BB; 3318—AA, AD).

Vouchers: *Lewis* 1060 (SAM); *Barker* 10649 (NBG); *Leighton* 599 (BOL); *De Vos* 1773 (STE); *Marloth* 8027.

Two forms occur: on sandy areas robust plants with longer styles and brown markings on the reverse of the perianth; and on more clayey ground smaller plants with shorter styles and often without brown markings on the perianth.

17. **Romulea barkerae** *De Vos* in Jl S. Afr. Bot. Suppl. 9: 106, fig. 25 (1972). Type: Cape, Vredenburg, Cape Columbine, *Barker* in NBG 273/67 (NBG, holo.!).

Closely related to *R. flava* (no. 15) from which it differs as follows: Plants 120—200 mm long. *Stem* short and hidden or up to 25 mm long. *Basal leaf* more or less 3-angled, almost T-shaped in transverse section, with only two wide stomatiferous grooves, 110—200 × 1,5—2,5 mm, minutely ciliate on the rib margins. *Inner bract* wholly membranous, sometimes with reddish veins. *Flowers* 25—38 mm long, white, each segment with a large black, yellow-margined blotch in the throat, outer segments bright green on the backs. *Style* 10—12 mm; stigmas reaching halfway up the anthers or to the tips.

Found only on the west coast near Cape Columbine (3217—DD).

Voucher: *Barker* in NBG 273/67.

Readily distinguished by its white flowers with a dark, yellow-margined blotch on each perianth segment in the throat.

18. **Romulea citrina** *Bak.*, Handb. Irid. 100 (1892); in F.C. 6: 38 (1896); Klatt in Dur. & Schinz, Consp. Fl. Afr. 5: 163 (1895); Bég. in Malpighia 23: 106 (1909); De Vos in Jl S. Afr. Bot. Suppl. 9: 111, figs 11 & 27 (1972). Type: Cape, Namaqualand near Modderfontein, *Bolus* 6619 (K, holo.!: BOL!; GRA!).

Plants 100—350(—450) mm long. *Corm* with a crescent-shaped basal ridge. *Stem* short or up to 200 mm long. *Basal leaves* 2 or more, filiform or compressed cylindrical, 100—450 × 0,7—1,5 mm, grooves narrow. *Inner bract* often slightly shorter than the outer, green with wide brown-streaked or -edged membranous margins, tip green or slightly scarious. *Flowers* 22—40 mm long, lemon-yellow, outer segments greenish or brownish on the backs. *Perianth tube* 4—7 mm long; segments oblanceolate or elliptical, 20—32 × (5—)8—10 mm. *Filaments* 5—8 mm; anthers 4—7 mm. *Style* 10—15 mm; stigmas more or less at the anther tips. *Capsules* shortly cylindrical, on arcuate peduncles which straighten later. *Chromosome no.* 2n=24.

Widespread in Namaqualand at low and higher altitudes (Kamiesberg), on sandy or stony ground. Map 6.

Vouchers: *Schlechter* 11121 (BOL; PRE; GRA; K; S; Z); *Acocks* 19440 (PRE; K; M); *Leipoldt* 3542 (BOL; PRE); *De Vos* 2278 (STE); *Barker* 3676 (NBG; BOL).

Distinguished by its corm, uniformly pale yellow flowers, two basal leaves in long-stemmed plants, and largely submembranous inner bracts.

19. **Romulea pearsonii** *De Vos* in Jl S. Afr. Bot. Suppl. 9: 113, figs 12 & 28 (1972). Type: Cape, Namaqualand, Kamiesberg, Khamsoap Ravine, *Pearson sub P. Sladen Mem. Exp.* 6550 (BOL, holo.!; K!).

Plants 100−250 mm long. *Corm* with a crescent-shaped basal ridge. *Stem* short, usually hidden by leaf bases, elongating to 100 mm in fruiting specimens. *Leaves* mainly basal, filiform, 100−250 × up to 1 mm, usually curved, grooves narrow. *Bracts* green, outer with narrow, and inner with wide brown-streaked membranous margins and tips. *Flowers* 30−50 mm long, lemon-yellow, outer segments on the backs yellowish green or brownish, or dark-veined. *Perianth tube* 4−5 mm long, almost cup-shaped; segments 25−40 × 8−14 mm, the outer slightly longer and narrower than the inner. *Filaments* 6−8 mm; anthers 7−10 mm long. *Style* 12−18 mm; stigmas at or below the anther tips. *Capsules* shortly cylindrical, on almost straight or slightly curved peduncles. *Chromosome no.* 2n = c. 24.

From Namaqualand on the Kamiesberg and flats to the west, on sandy soil (3017−BB; 3018−AB, AC).

Vouchers: *Acocks* 19466 (PRE; K; M); *Oliver* 3529, 3532 (STE); *Schweickerdt* 2550 (PRE; K); *De Vos* 1616 (STE); *Rourke* s.n. 18.7.68 (NBG).

Distinguished from *R. citrina* (no. 18) by its bracts with pronounced brown streaks on the margins and tip, and from *R. luteoflora* (no. 49) by its corm and its ripe capsules which are on almost straight peduncles on an elongated stem.

20. **Romulea neglecta** (*Schultes*) *De Vos*, comb. nov., excl. syn. Andr. (1801) and Ker-Gawl. (1805 & 1809).

Ixia neglecta Schultes, Syst. Veg. Mant. 1: 279 (1822), partly, excl. syn. Andr. (1801) and Ker-Gawl. (1805 & 1809). Iconotype: Curtis's bot. Mag. 36: t.1476 (1812).

Trichonema speciosum Ker-Gawl. in Curtis's bot. Mag. 36: t.1476 (1812), excl. syn., non Ker-Gawl. (1805). *R. speciosa* (Ker-Gawl.) Bak. in J. Linn. Soc., Bot. 16: 89 (1877), partly, excl. syn. Andr. (1801) and Ker-Gawl. (1805 & 1809) and Eckl. (1827). Iconotype: as above.

R. oliveri De Vos in Jl S. Afr. Bot. Suppl. 9: 116 (1972). Type: Cape, Namaqualand, farm Welkom SE of Kamiesberg peak, *Oliver* 3169 (STE, holo.!; PRE!).

Plants 150-300 mm long. *Corm* with a crescent-shaped basal ridge. *Stem* short, hidden or up to c. 150 mm long. *Basal leaves* 2 or more, terete, filiform, c. 150−300 × 1 mm, recurved, grooves narrow. *Bracts* green, outer 17−30 mm long, densely veined, often with a stronger median vein, margins narrow, brownish; inner with wide, brown or brown-streaked membranous margins and tip. *Flowers* 35−45 mm long, bright rosy-magenta, cup striped purple and yellow, outer segments on the backs yellow with 5 purple veins and feathered veining. *Perianth tube* 4−5 mm long; segments 25−35 × c. 10 mm subobtuse, minutely emarginate. *Filaments* 4 mm; anthers 8-10 mm long, pale yellow. *Style* 10−12 mm; stigmas below the anther tips.

Found on the Kamiesberg, Namaqualand, on a moist sandy slope (3018−AC).

Voucher: *Oliver* 3169 (STE; PRE).

This species is nearest the sympatric *R. pearsonii* (no. 19) from which it differs in flower colouring, larger bracts, the outer with a green tip, and anthers which are longer than the filaments. Leaf structure as in *R. pearsonii*. Schultes's description of *Ixia neglecta* is identical with Ker-Gawler's description of *Trichonema speciosum* (1812) and does not fit Andrew's figure of *I. bulbocodium* var. *speciosissima* which Schultes, however, cited as a synonym.

21. **Romulea hirsuta** (*Eckl. ex Klatt*) *Bak.* in J. Linn. Soc., Bot. 16: 89 (1877); Handb. Irid. 102 (1892); F. C. 6: 40 (1896); Eckl., Top. Verz. 19 (1827), nom. nud.; Klatt in Abh. Naturforsch. Ges. Halle 15: 398 (1882); Bég. in Malpighia 23: 88 (1909); G. J. Lewis in Adams. & Salter, Fl. Cape Penins. 221 (1950); De Vos in Jl S. Afr. Bot. Suppl. 9: 125, figs 30 & 37 (1972). Syntypes: Cape, *Bergius* s.n. (not found); *Ecklon & Zeyher* 207 (B†; SAM, lecto.!).

Plants 60−300 mm long. *Corm* bell-shaped, with fine fibrils on a circular basal ridge. *Stem* short or up to 180 mm long. *Basal leaves* 2 or more, filiform or compressed cylindrical, 50−300 × 0,5−4 mm, glabrous or minutely ciliate on the rib margins, grooves narrow or wide. *Bracts*

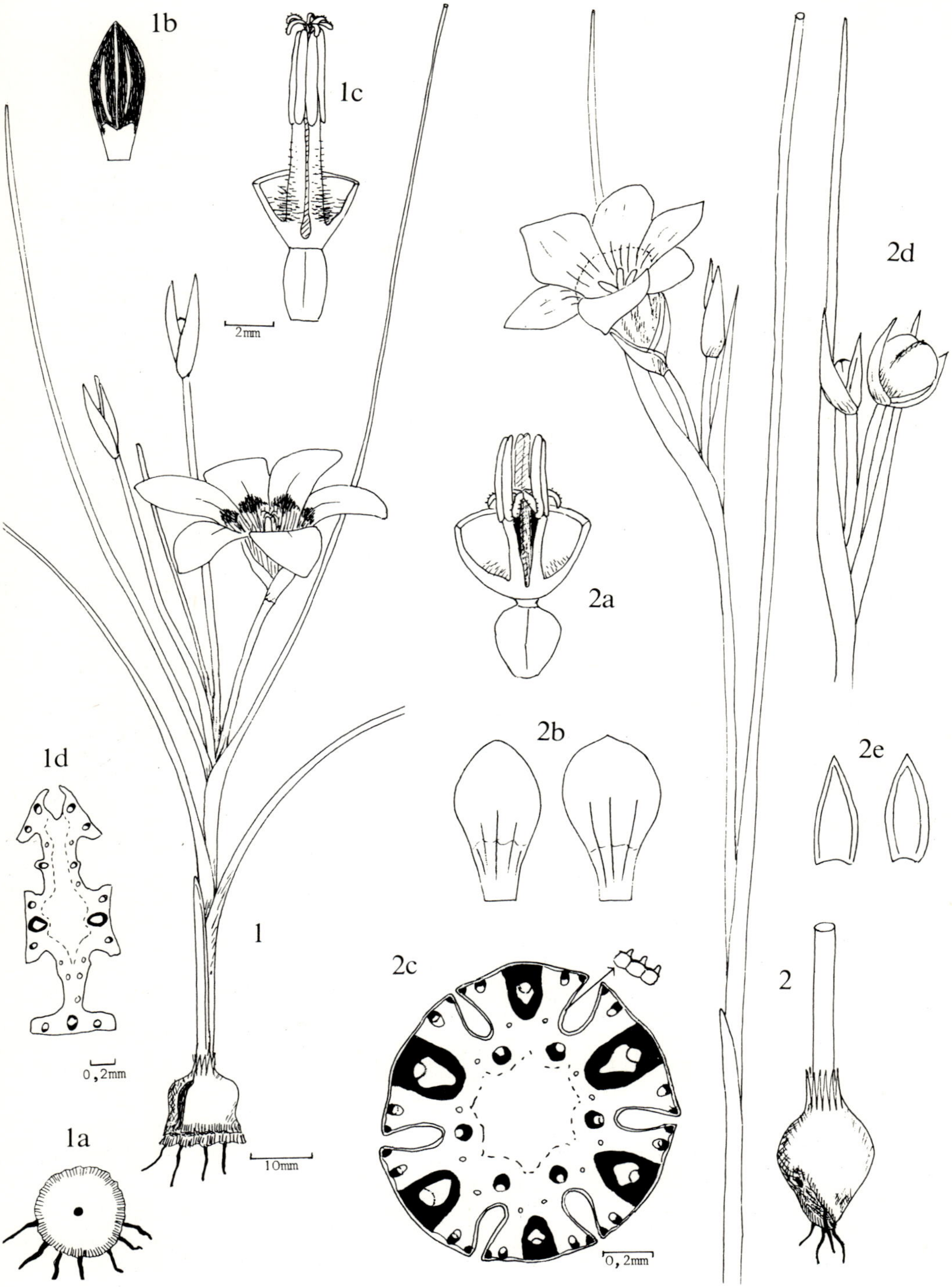

1b
1c
2mm
2d
2a
1
1d
2b
2e
0,2mm
2c
1a
2
10mm
0,2mm

green, inner with wide or narrow, brown or colourless membranous margins. *Flowers* (15−)20−45 mm long, apricot-pink to dark old-rose or sometimes rosy-magenta, often with dark brownish red or purplish black blotches in the throat, cup golden or orange-yellow or sometimes paler. *Perianth tube* 3−6 mm long, funnel-shaped to almost cup-shaped; segments 15−35 × 5−12 mm. *Filaments* 4−8 mm; anthers 3−7 mm. *Style* 8−16 mm; stigmas at or near the anther tips. *Capsules* ellipsoid, on suberect or slightly spreading peduncles. *Chromosome no.* 2n=24. Fig. 7: 1.

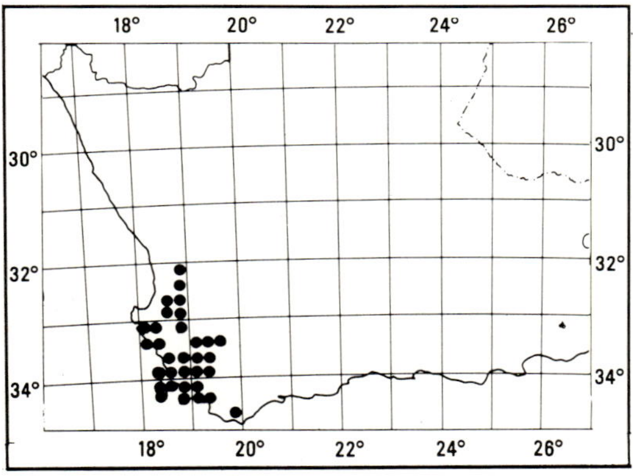

MAP 7.— **Romulea hirsuta**

Widespread in south-western Cape districts from Clanwilliam to Worcester, to the Cape Peninsula and Bredasdorp. Map 7.

A large polymorphic species with four varieties:

1 Perianth without dark blotches in the throat:

  2 Flowers pink in the upper part .... (a) var. *hirsuta*

  2 Flowers salmon or 'copper' in the upper part
  ....(c) var. *cuprea*

1 Perianth with dark blotches in the throat:

  2 Perianth segments generally less than 12 mm wide:

    3 Widest leaf blade c. 1 mm in diam. ....
    ....(a) var. *hirsuta*

    3 Widest leaf blade 1,5−4 mm in diam. ....
    .... (b) var. *zeyheri*

  2 Perianth segments 12−20 mm
  wide ....(d) var. *framesii*

(a) var. **hirsuta.**

Trichonema hirsutum Steud. ex Klatt in Linnaea 34: 665 (1865−66); Steud., Nom. Bot. edn 2, 2: 702 (1841), nom. nud.

Ixia campanulata Lam., Tabl. Encycl. 1: 109 (1791), non Houtt. (1780). *?Bulbocodium campanulatum* Kuntze, Rev. Gen. 2: 700 (1809).

Ixia filifolia var. *B* F. Delaroche in Red., Liliac. 5 sub t. 25 (1809); Poir. in Lam., Encycl. Suppl. 3: 201 (1813), partly.

R. ramosa Eckl., Top. Verz. 19 (1827), nom. nud. *Trichonema ramosum* Steud. 2: 702 (1841), nom. nud.

R. uncinata Klatt in Abh. Naturforsch. Ges. Halle 15: 401 (1882); Bak. in F. C. 6: 41 (1896), pro syn. Type: Cape, C. B. S., *Pappe* s.n. (B?); *Pappe* in SAM 20696 neo.!).

R. rubrolutea Bak. in Kew Bull. 1906: 25 (1906). Syntypes: Cape, Clanwilliam, Olifants River, *Penther* 678 (K, lecto.!); Piketberg, *Penther* 633 (K?; BM?).

Stem short or up to 90 mm long. *Leaves* filiform, c. 1 mm in diam., glabrous or minutely ciliate, grooves usually narrow. *Inner bract* with brown membranous margins. *Flowers* 20−40 mm long, dark rosy-pink or apricot-pink, or rosy-magenta, usually with dark blotches in the throat, cup golden-yellow or with very little yellow. *Perianth segments* (5−)6−10 mm wide.

From Clanwilliam to Bredasdorp, up to 900 m altitude, in sandy soil.

Vouchers: *Salter* 3581 (BOL; K); *MacOwan* 2565 (SAM; K); *De Vos* 2048 (STE); *Thompson* 2600 (STE).

(b) var. **zeyheri** *(Bak.) De Vos* in Jl S. Afr. Bot. Suppl. 9: 129 (1972). Type: Cape, ex Herb. Zeyheri s.n. sub *Geissorhiza zeyheri* Spreng. (K, holo.!).

R. rosea Eckl. var. *zeyheri* Bak., Handb. Irid. 103 (1892). *R. zeyheri* (Bak.) Bég. in Bot. Jb. 38: 338 (1907); in Malpighia 23: 105 (1909), non Eckl. 1827.

R. klattii Bég. in Bot. Jb. 38: 333 (1907); in Malpighia 23: 92 (1909). Syntypes: Cape, Malmesbury, near Hopefield, *Bachmann* 1579 (B!; Z!); Darling, *Bachmann* 513 (B!).

R. bulbocodioides var. *ambigua* Bég. in Annu. Conserv. Jard. bot. Genève 11−12: 163 (1908); in Malpighia 23: 110 (1909). Type: Cape, *Drège*, Irid. 208 (G, lecto.!).

Stem usually long, up to 180 mm, often branched. *Leaves* subterete or compressed

---

FIG. 7.−1, **Romulea hirsuta**, habit, × 7/8; 1a, base of corm; 1b, outer perianth segment, lower face; 1c, perianth tube, stamens and pistil; 1d, transverse section of leaf of var. **zeyheri** (*De Vos* 1098). 2, **R. aquatica**, habit, × 1; 2a, perianth tube, stamens and pistil; 2b, outer (left) and inner (right) perianth segments; 2c, transverse section of leaf; 2d, almost mature capsules; 2e, outer (left) and inner (right) bract (*De Vos* 1738).

cylindrical, 1,5—4 mm wide, glabrous or sometimes ciliate, grooves wide. *Inner bract* with wide colourless membranous margins. *Flowers* 30—45 mm long, dark rosy pink or dark apricot-pink, usually with dark blotches in the throat, cup often orange-yellow. *Perianth segments* 8—12 mm wide. Fig 7: 1d.

From Vredenburg to Paarl, at low altitudes, in sandy soil.

Vouchers: *Barker* 10399 (NBG; STE); *De Vos* 2016 (STE); *Salter* 3007 (BOL, partly in K); *Lewis* 5981 (NBG).

Intermediates between this variety and var. *hirsuta* have been found.

(c) var. **cuprea** *(Bég.) De Vos* in Jl S. Afr. Bot. Suppl. 9: 131 (1972). Iconotype: *Herbert* MS. in Lindley Library, R. Hort. Soc., London, sub *Trichonema cupreum*.

R. cuprea Herb. ex Bak. in J. Bot., Lond. 5: 236 (1876); in J. Linn. Soc., Bot. 16: 89 (1877); in F. C. 6: 42 (1896); Klatt in Abh. Naturforsch. Ges. Halle 15: 402 (1882). *Bulbocodium cupreum* (Bak.) Kuntze, Rev. Gen. 2: 700 (1891). *R. rosea* var. *cuprea* (Bak.) Bég. in Malpighia 23: 63 (1909).

Very closely related to var. *hirsuta*, differing as follows: *Inner bract* with white membranous margins. *Flowers* pale apricot, without dark blotches.

From Clanwilliam to Bredasdorp.

Vouchers: *Salter* 4663 (BOL; K); *Marsh* 894 (STE); *De Vos* 1775 (STE); *Bolus* 3760, partly (BOL; PRE).

(d) var. **framesii** *(L. Bol.) De Vos* in Jl S. Afr. Bot. Suppl. 9: 132 (1972). Type: Cape, Malmesbury, near Darling, *Ross-Frames* in BOL, 18993 (BOL, holo.!; K!).

R. framesii L. Bol. in J. Bot., Lond. 69: 13 (1931).

Stem usually long, up to 140 mm. *Leaves* subterete or compressed cylindrical, 1—2 mm wide, sometimes minutely ciliate, grooves wide or narrow. *Inner bract* with brown-edged membranous margins, sometimes slightly shorter than the outer. *Flowers* 30—50 mm long, dark old-rose, each segments with 3 subequal blotches in the throat, or with the median black-purple blotch wider than the pale lateral ones. *Perianth segments* usually 12—20 mm wide, the outer irregularly marked with pink and green on the backs.

From the Malmesbury district.

Vouchers: *Bolus* in BOL 20721 (BOL; K); *Ross-Frames* in BOL 18993 (BOL); *Penberthy* in NBG 2832/35 (BOL).

22. **Romulea triflora** *(Burm. f.) N. E. Br.* in Kew Bull. 1929: 131 (1929); G. J. Lewis in Adamson & Salter, Fl. Cape Penins. 222 (1950); De Vos in Jl S. Afr. Bot. Suppl. 9: 122, fig. 36 (1972).

Crocus triflorus Burm. f., Prodr. Fl. Cap. 2 (1768). Type: S. coll., s.n. in Burman herb. sub *Crocus triflorus* (G, holo.!).

Ixia filifolia var. *A* F. Delaroche in Red., Liliac. 5: t. 251, fig. 2 (1809); Poir. in Lam., Encycl. Suppl. 3: 201 (1813), partly. *Trichonema filifolium* (F. Delaroche) Ker-Gawl., Irid. Gen. 82 (1827). *R. filifolia* (F. Delaroche) Eckl., Top. Verz. 20 (1827). *Bulbocodium filifolium* (F. Delaroche) Kuntze, Rev. Gen. 2: 700 (1891). Iconotype: Red., Liliac. 5: t. 251, fig. 2 (1809).

Ixia crocea Thunb., Fl. Cap. 1: 218 (1811). Type: Cape, C. B. S., *Thunberg* s.n. herb. no. 946 (UPS, holo.!).

?*Ixia sublutea* Lam., Encycl. 3: 335 (1789), e descr. *Geissorhiza sublutea* (Lam.) Ker-Gawl. in König & Sims, Ann. Bot. 1: 224 (1805), non (1803). *R. sublutea* (Lam.) Bak. in J. Linn. Soc., Bot. 16: 88 (1877); Handb. Irid. 100 (1892); in F. C. 6: 37 (1896), excl. syn. *R. aurea*; Klatt in Abh. Naturforsch. Ges. Halle 15: 399 (1882); Bég. in Malpighia 23: 98 (1909), excl. syn. *Trichonema roseum* (L.) Ker-Gawl.; N. E. Br. in J. Linn. Soc., Bot. 48: 16 (1928). Type not found.

R. schlechteriana Schinz in Bull. Herb. Boissier 3: 395 (1895). Type: Cape, Rondebosch, *Schlechter* 852 (Z, holo.!).

Closely related to *R. hirsuta* (no. 21) from which it differs as follows: *Leaves* filiform, glabrous, 0,5—1 mm diam., grooves narrow. *Inner* bract with wide, colourless or brown-edged membranous margins. *Flowers* bright golden yellow, sometimes with a diffuse brown zone in the throat, rarely white with a yellow cup. *Perianth tube*

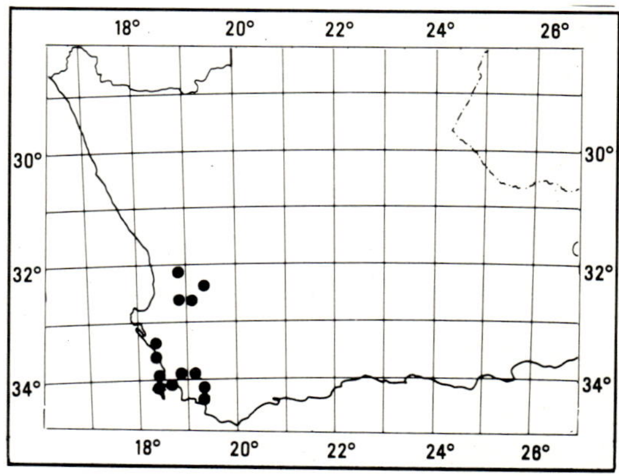

MAP 8.— **Romulea triflora**

widely funnel-shaped. *Filaments* 4—5 mm long. *Chromosome no.* 2n×24.

Widely distributed in south-western Cape districts from Clanwilliam to the Cape Peninsula and to Caledon, at low altitude, on sandy loam. Map 8.

Vouchers: *Boucher* 1593 (STE); *De Vos* 1918 (STE); *Hutchinson* 659 (BOL; PRE; K); *Esterhuysen* 18868 (BOL); *Compton* 13771 (NBG).

Distinguished by its bell-shaped corm and bright golden yellow or rarely white flowers, without any red colouring. Some herbarium specimens are not easily distinguished from *R. hirsuta* var. *cuprea* (no. 21c) which, however, has some anthocyanin in its perianth.

23. **Romulea gracillima** *Bak.,* Handb. Irid. 103 (1892); in F. C. 6: 41 (1896); Bég. in Malpighia 23: 73 (1909); De Vos in Jl S. Afr. Bot. Suppl. 9: 133, figs 33 & 38 (1972). Type: Cape, Drakenstein Mtns, *Drège* s.n. *(Trichonema cruciatum, a)* (K, holo.!; BM!; G!; P!; OXF!).

Closely related to *R. hirsuta* (no.21) from which it differs in the following: *Stem* up to 100 mm. *Leaves* filiform, generally less than 1 mm in diam., glabrous, grooves very narrow. *Inner bract* with narrow colourless membranous margins. *Flowers* 15—25 mm long, pale pink, sometimes with red lines in the throat. *Perianth tube* 3 mm long; segments 12—18 × 3—6 mm. *Filaments* 3—5 mm; anthers 2—4 mm long. *Style* 7—8 mm.

On several south-western Cape mountain ranges from the Cape Peninsula to Stellenbosch and to Bredasdorp. Map 9.

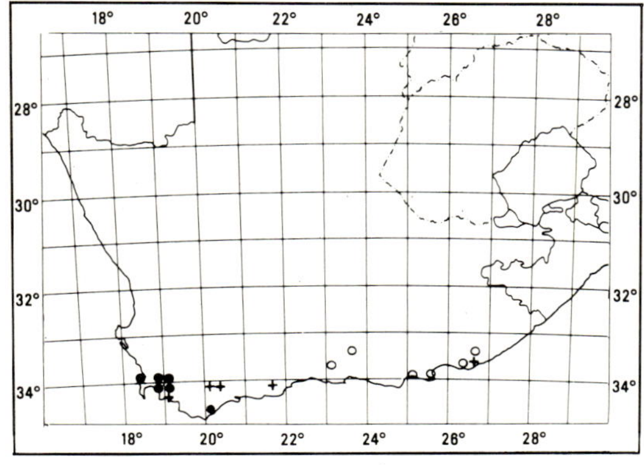

MAP 9.—● **Romulea gracillima**
    ○ **R. pratensis**
    + **R. gigantea**

Vouchers: *Lewis* 1664 (SAM); *Esterhuysen* 17582, 32181 (BOL); *De Vos* 2138, 1976 (STE).

May perhaps be regarded as a small montane variety of *R. hirsuta* (no. 21), with smaller, paler flowers and narrower perianth segments.

24. **Romulea sladenii** *De Vos* in Jl S. Afr. Bot. Suppl. 9: 135, figs 31 and 39 (1972). Type: Cape, Vanrhynsdorp, plateau on the Matsikamma Mtn, *De Vos* 2018 (STE, holo.!).

Related to *R. hirsuta* (no. 21) from which it differs as follows: *Corm* with a wider circular basal ridge wider than the corm itself; with tunics irregularly lacerated on the ridge into irregular fibril groups. *Basal leaves* filiform, 50—300 mm long, generally less than 1 mm in diam., glabrous, with narrow grooves. *Inner bract* with wide, almost colourless or brown-edged membranous margins. *Flowers* 22—33 mm long, white with a bright yellow cup, outer segments on the backs reddish or sometimes greenish. *Perianth tube* 4—5 mm long; segments 15—25 × 4—9 mm. *Filaments* 5—7 mm, orange-yellow; anthers 4—6 mm long, pale yellow. *Style* c. 13 mm. *Capsules* usually two, on widely patent, horizontal peduncles.

Found only on the Gifberg plateau, Vanrhynsdorp (3118—DC).

Vouchers: *Phillips sub P. Sladen Mem. Exp.* 7506 (BOL; SAM; K); *Compton* 20846 (NBG); *De Vos* 2018 (STE).

A montane species recognized by its corm, white flowers with the outer segments reddish or greenish, and capsules on widely patent, horizontal peduncles.

25. **Romulea tortilis** *Bak.* in Bull. Herb. Boissier sér. 2, 4: 1003 (1904); Bég. in Malpighia 23: 95 (1909); De Vos in Jl S. Afr. Bot. Suppl. 9: 132 (1972). Type: Cape, near Porterville, *Schlechter* 4890 (Z, holo!; GRA!; PRE!; SAM!; B!; G!; K, partly!).

Closely related to *R. hirsuta* (no. 21) from which it differs as follows: *Plants* up to 120 mm long. *Corm* with minute parallel fibrils or irregular scales on the circular basal ridge. *Stem* up to 25 mm. *Basal leaves* filiform, more or less spirally twisted, up to 1 mm in diam. *Inner bract* with wide brown or brown-edged membranous margins. *Flowers* up to 30 mm long, reddish or dark old-rose or magenta, often with dark blotches in the throat, sometimes with a yellow cup. *Perianth tube* to 4 mm long; segments 15—25 × 5—9 mm. *Style* 10—12 mm, with branches sometimes repeatedly branched; stigmas 6 or more.

From the western Cape districts of Clanwilliam and Piketberg (3218—BB, BD, DB; 3318—BB).

Two varieties are recognized, differing in the number of stigmas.

(a) var. **tortilis.**

R. torta Bak. in Kew Bull. 1906: 24 (1906). Syntypes: Cape, Olifants River, Clanwilliam, *Penther* 687 (K, lecto.!; BOL!; BM!; M!); *Penther* 624.

Flowers old-rose or reddish, with a dark red or purple blotch on each segment in the throat, cup golden-yellow. *Style* with bifid branches; stigmas 6.

Vouchers: *Schlechter* 4890, 10735 (BOL; GRA; PRE; K; G; S; US); *Penther* 687 (BOL).

(b) var. **dissecta** *De Vos* in Jl S. Afr. Bot. Suppl. 9: 133 (1972). Type: Cape, Clanwilliam near The Rest, *Bolus* 21265 (BOL; holo.!).

Flowers magenta or reddish purple, often with dark blotches in the throat, with very little yellow in the cup. *Style* branches multifid; stigmas more than 6.

Vouchers: *Leipoldt* 3824 (BOL); *Gillett* 3713 (BOL); *Bolus* 23190 (BOL).

26. **Romulea multisulcata** *De Vos* in Jl S. Afr. Bot. Suppl. 9: 139, fig. 40 (1972). Type: Cape, Vanrhynsdorp, between Vanrhyn's Pass and Nieuwoudtville, *De Vos* 2183 (STE, holo.!).

Plants aquatic, with corm and base of stem submerged, 300—500 mm long. *Corm* with a high crescent-shaped basal ridge. *Stem* 60—220 mm long, extending above water level. *Basal leaves* 2, terete, erect, 300—500 × 1—2 mm, with 6—8 narrow grooves. *Bracts* greenish or reddish green, outer with narrow, inner with wide colourless or fawn membranous margins. *Flowers* 25—32 mm long, buttercup-yellow, outer perianth segments with a median brown zone on the backs. *Perianth tube* 3—5 mm long, funnel-shaped; segments 15—25 mm long, outer c. 7 mm wide, inner 9 mm wide. *Filaments* 4—6 mm; anthers 6—8 mm long. *Style* 12—15 mm; stigmas at or just above the anther tips. *Capsules* subglobose, 3-lobed, on widely patent peduncles. *Chromosome no.* 2n=24.

Found only in seasonal pools which are dry in summer, on the plateau of the Bokkeveld Mountains east of Vanrhynsdorp (3119—AC, CA).

Vouchers: *Loubser* 944 (NBG); *Markötter* STE 18966 (STE); *De Vos* 2184 (STE).

Distinguished by its aquatic habit, leaves with more than the usual four stomatiferous grooves, yellow flowers with the inner perianth segments widely spathulate, and globose capsules.

27. **Romulea aquatica** *G. J. Lewis* in Jl S. Afr. Bot. 4: 8 (1938); De Vos in Jl S. Afr. Bot. Suppl. 9: 141, figs 32 & 41 (1972). Syntypes: Cape, Elandsvlei north of Piketberg, *Barker* 190 (NBG, lecto.!); between Hopefield and Moorreesburg, *Salter* 3880 (BOL!; K!); *Mathews* in BOL 22169!.

Related to *R. multisulcata* (no. 26) from which it differs as follows: *Plants* up to 600 mm long. *Stem* 120—350 mm long, often deflexed near the top in the fruiting stage. *Basal leaf* single, 0,8—1,5 mm in diam., with 5—8 narrow grooves. *Bracts* green or greenish, inner with wide, colourless or brown-speckled membranous margins. *Flowers* slightly smaller, white or cream in the upper half, buttercup-yellow in the lower half, fragrant. *Perianth tube* 4 mm long, cup-shaped; segments obtuse, subobtuse or emarginate, outer 6—9 mm wide, inner 9—13 mm wide. *Filaments* 3—4 mm; anthers 3—5 mm long, slightly spreading. *Style* 4—8 mm; stigmas not reaching the anther tips. *Capsules* on erect peduncles. *Chromosome no.* 2n=24. Fig. 7: 2.

In seasonal pools in western Cape districts from Piketberg to Malmesbury (3218—DD; 3318—AB, BA).

Vouchers: *Lewis* 149 (SAM); *Barker* 7391 (NBG); *Salter* 3880 (BOL, K); *De Vos* 1622, 2015 (STE).

Readily distinguished by its aquatic habit, flower colour and fragrance, inner perianth segments wider than the outer, short style, and globose capsules which later become so heavy that the stem bends near its top.

28. **Romulea minutiflora** *Klatt* in Abh. Naturforsch. Ges. Halle 15: 339 (1882); in Dur. & Schinz, Consp. Fl. Afr. 5: 165 (1895); Bak., Handb. Irid. 102 (1892); in F.C. 6: 40 (1896); Bég. in Malpighia 23: 79 (1909); G. J. Lewis in Adams. & Salter, Fl. Cape Penins. 222 (1950); De Vos in Jl S. Afr. Bot. Suppl. 9: 146, figs 42 & 60 (1972). Type: Cape, Worcester, versus Hexflussberg, *Drège 538* (B, holo.†; S, lecto.!).

Plants small, 60—200 mm long. *Corm* usually somewhat elongated vertically, with a high crescent-shaped basal ridge. *Stem* short, hidden. *Leaves* basal, often distichous, filiform to compressed cylindrical, 60—200 × 0,5—1,5 mm, often arcuate, grooves narrow. *Bracts* submembranous or greenish upwards; inner bract almost wholly membranous, with wide, usually brown-spotted membranous margins. *Flowers* 7—15 mm long, pale mauve or lilac, rarely whitish, often with a violet circle in the throat, cup greenish yellow, outer segments greenish or mottled on the backs. *Perianth tube* to 3,5 long; segments 4—9 × c. 2 mm. *Filaments* 2—4 mm; anthers 1,5—2 mm. *Style* 4—6 mm. *Capsules* ellipsoid, up to 15 mm long, on arcuate peduncles which straighten later. *Chromosome no.* 2n=26.

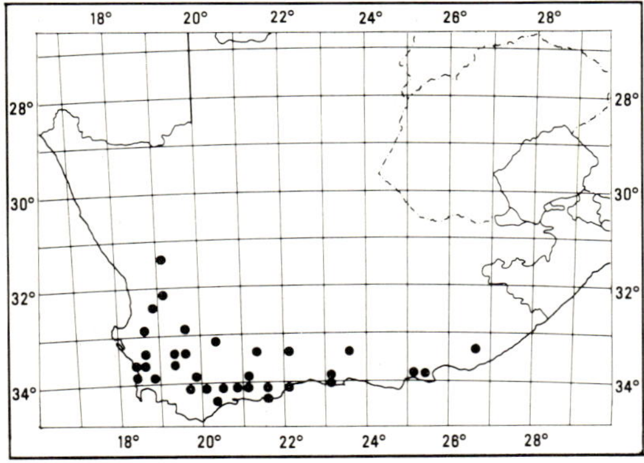

MAP 10.— **Romulea minutiflora**

Found in western and southern Cape districts from Calvinia to the Cape Peninsula and to Grahamstown, the Little Karoo and south-western Karoo; also introduced into Australia. Map 10.

Vouchers: *Schlechter* 8889 (BOL; PRE; GRA; K; S: Z); *Salter* 3546 (BOL; K); *Jacot Guillarmod* 5009 (NBG; RUH; STE, partly); *De Vos* 2234 (STE); *Barker* 7364 (NBG).

One of the most widespread *Romulea* species, distinguished by its small pale flowers (smallest amongst the Cape species), corm with a high basal ridge and inner bract usually with relatively large brown spots.

29. **Romulea sinispinosensis** *De Vos* in Jl S. Afr. Bot. Suppl. 9: 147, fig. 43 (1972). Type: Cape, Vredendal, Doringbaai near coast, *De Vos* 2106 (STE, holo.!).

Very closely related to *R. minutiflora* (no. 28) from which it differs as follows:

Plants 120—200 mm long. *Leaves* 120—200 × c. 1 mm. *Inner bract* largely membranous with slender brown veins in the centre and white membranous margins with faint brownish spots towards the tip. *Flowers* 16—20 mm long, cream or white, yellowish green in the lower half and on the backs of the outer segments. *Perianth segments* 10—12 × 3—4 mm. *Anthers* 3,5—4 mm long. *Style* 6—8 mm long. *Chromosome no.* 2n=50.

Found only once in the western Cape district of Vredendal near the coast at Doringbaai (3118—CC).

Voucher: *De Vos* 2106 (STE).

Might be considered as a slightly larger form of *R. minutiflora* (no. 28) distinguished by slightly larger flowers with slightly wider perianth segments, longer anthers and style, and by the faint marks on the margins of the inner bracts. The morphological features and chromosome number indicate a possible amphidiploid origin from *R. minutiflora* and some other species of *Romulea*.

30. **Romulea pratensis** *De Vos* in Jl S. Afr. Bot. Suppl. 9: 198, figs 62 & 66 (1972). Type: Cape, near and in Grahamstown, *Dyer* 1619 (GRA, holo.!; PRE!).

Plants 120—250 mm long. *Corm* with a small, often high, crescent-shaped basal ridge and a row of parallel fibrils or teeth very sharply bent over the ridge into fibrils minutely forked at their ends. *Stem* short, hidden. *Leaves* basal, filiform to compressed cylindrical, 120—250 × 1—2 mm, grooves narrow. *Bracts* green in the upper half, submembranous in the lower, the inner with wide brown-speckled or colourless membranous margins. *Flowers* (12—) 15—20 (—24) mm long, pale rose or lilac-rose to almost white, often with 1—3 dark lines in the throat, outer segments brownish purple or green with dark lines on the backs. *Perianth tube* 2—3,5 mm long; segments 8—15 × 3—5 mm. *Filaments* 3—4 mm, sometimes unequal in length; anthers 3—4 mm long. *Style* 6—8 mm; stigmas often reaching to the anther tips. *Capsules* on curved peduncles which straighten later. *Chromosome no.* 2n=44. Fig. 8:1.

From the south-eastern Cape districts of Uitenhage, Grahamstown. Port Elizabeth and Alexandria, in grassveld; also at the top of Prince Alfred's Pass, Uniondale. Map 9.

Vouchers: *Fries, Norlindh & Weimarck* 1053 (PRE; K; S); *Jacot Guillarmod* 5241 (STE; GRA); *De Vos* 1734, 1735 (STE); *Martin* in STE 30195 (STE); *Dyer* 1619 (PRE; GRA).

1a
1b
1
2b
2c
2d
2e
2a
2
3mm
2mm
0,2mm

In morphological features this species shows relationship with both *R. minutiflora* (no. 28) and *R. rosea* var. *australis* (no. 59c), the corm, especially, being intermediate between the two. It might well have originated from these as an amphidiploid, as is indicated by its chromosome number of 2n=44. It has often been mistaken for *R. rosea* var. *australis*, but it differs from the latter in its corm and somewhat greener bracts.

31. Romulea gigantea *Bég.* in Bot. Jb. 38: 333 (1907); in Malpighia 23: 76 (1909); Martin & Noel in Publ. Dept. Bot. Rhodes Univ. 30 (1960); De Vos in Jl S. Afr. Bot. Suppl. 9: 117, fig. 29 (1972). Type: Cape, Riversdale, *Rust* 622 (B, holo.!).

Related to *R. pratensis* (no. 30) from which it differs as follows: *Plants* 200−500 mm long. *Stem* short or elongated to 500 mm. *Basal leaves* 2 or more, compressed cylindrical, 200−500 × 1−3 mm. *Bracts* green, the inner usually shorter than the outer, with wide brown-edged membranous margins and a green tip. *Flowers* white, bluish white or pale lilac, outer segments on the backs with irregular green and purplish brown marks or three longitudinal stripes. *Filaments* 3−7 mm long. *Chromosome no.* 2n=c. 42 or 44. Fig. 8: 2.

From southern Cape coastal districts from Caledon to Riversdale, and in the Bathurst district, in moist localities. Map 9.

Vouchers: *Fries, Norlindh & Weimarck* 1369 (PRE; SAM; K; LD; S); *Britten* 770 (GRA; PRE); *Bain* in RUH 3256; *Martin* in STE 30196; *De Vos* 1980 (STE).

The tunic fibres on the basal ridge of the corm, as well as the polyploid chromosome number of 2n=c. 42 or 44, might show that this species is, like *R. pratensis* (no. 30) intermediate between sections *Romulea* and *Roseae*. The small flowers resemble those of *R. rosea* var. *australis* (no. 59c) superficially, but *R. gigantea* differs from *R. rosea* in its corm, often elongated stem and green bracts. From *R. pratensis* it differs mainly in its often elongated stem and green bracts.

2. Section **Aggregatae**

Aggregatae *De Vos* in Jl S. Afr. Bot. Suppl. 9: 170 (1972). Type species: *R. setifolia* N.E.Br.

Corm with tunics split into a row of small clusters of minute fibrils on a crescent-shaped or circular basal ridge. *Stem* short and hidden or elongated. *Foliage leaves* all basal or 1−2 basal and a few cauline, with 4 grooves. *Bracts* green or sometimes submembranous, inner with wide membranous margins and usually a green tip. *Flowers* of various colours. *Perianth tube* short. *Capsules* ellipsoid, on erect or patent peduncles. *Chromosome no.* 2n=24, 30, 32, c. 54.

The eight species of this section have previously (De Vos, 1972) been placed in two subsections which differ mainly in corm shape and chromosome number.

Mainly from Ceres to East London, also in the Nieuwoudtville area (excluding the extreme south-western Cape Province).

32. Romulea amoena *Schltr. ex Bég.* in Bot. Jb. 38: 334 (1907); in Malpighia 23: 90 (1909); De Vos in Jl S. Afr. Bot. Suppl. 9: 194, fig. 59 (1972). Type: Cape, Calvinia, Onder-Bokkeveld, Papelfontein, *Schlechter* 10896 (G holo.!; BOL!; GRA!; PRE!; BM!; K; S!; Z!).

Plants 80−300 mm long. *Corm* bell-shaped, with minute fibril clusters on a circular basal ridge. *Stem* short, hidden, or up to 100 mm long. *Basal leaves* 2 or 3−4, filiform, 70−300 × c. 1 mm, grooves narrow. *Bracts* largely green or sometimes reddish; inner sometimes slightly shorter than the outer, with wide, brown-streaked or almost colourless membranous margins. *Flowers* 25−45 mm long, carmine-red to deep rosy-pink, with a large purple-black blotch on each segment in the throat, cup cream, with dark V-shaped markings near the base, outer segments with red and

---

FIG. 8.−1, **Romulea pratensis**, habit, × 1; 1a, outer perianth segment, lower face; 1b, outer (left) and inner (right) bract (*De Vos* 1734 & 1743). 2, **R. gigantea**, habit, × 1; 2a, perianth tube, stamens and pistil; 2b, outer perianth segment, lower face; 2c, outer (left) and inner (right) bract; 2d, mature and half mature capsules; 2e, transverse section of leaf (*De Vos* 1910).

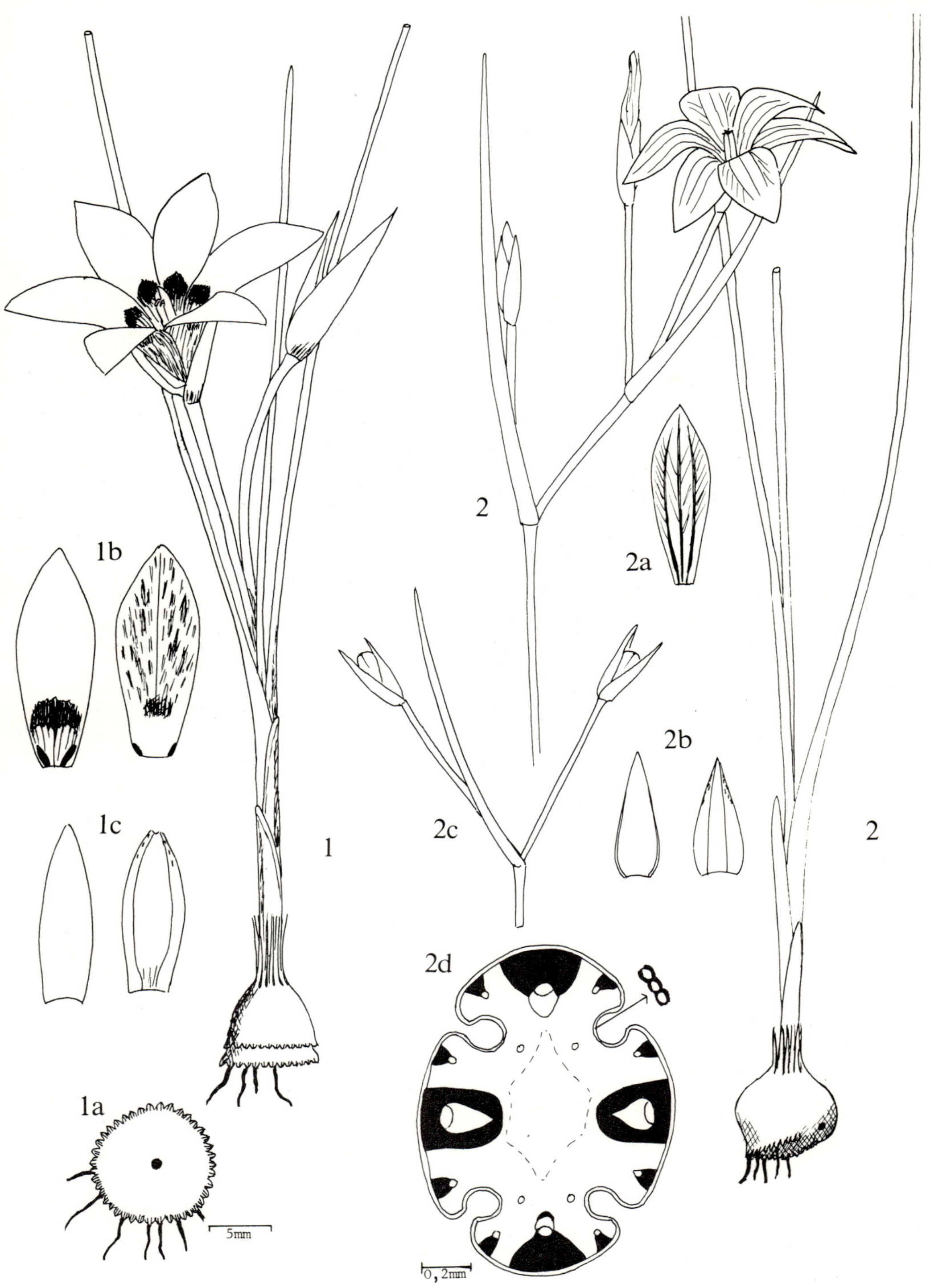

yellow irregular markings or purplish red. *Perianth tube* 5−7 mm long; segments subequal or the inner slightly shorter and wider, 18−35 × 9−15 mm. *Filaments* 3−5 mm, widened and pilose at the bases; anthers c. twice as long. *Style* 10−15 mm. *Chromosome no.* 2n=24. Fig. 9:1.

Found in the western Cape district of Calvinia on the Bokkeveld Mountains plateau near Nieuwoudtville (3119−AC).

Vouchers: *Burger* in STE 30198 (STE); *Lewis* in SAM 60232 (SAM; STE; K); *Hanekom* 3119.

Distinguished by its bell-shaped corm with minute fibril clusters on the circular basal ridge, red or deep pink flowers with dark blotches in the throat, and anthers about twice as long as the filaments.

33. **Romulea sanguinalis** *De Vos* in Jl S. Afr. Bot. Suppl. 9: 191, fig. 58 (1972). Type: Cape, Calvinia, Menzieskraal, 39 km from Nieuwoudtville, *Burger* in STE 30207 (STE, holo.!).

Very closely related to *R. amoena* (no. 32) from which it differs as follows: *Corm* with a crescent-shaped basal ridge forming an almost complete circle, with minute fibril groups on the ridge. *Stem* usually shortly extended, 40−80 mm long. *Basal leaves* 2, 200−350 × 0,7 mm. *Flowers* carmine-red without any markings inside. *Perianth segments* 8−12 mm wide. *Filaments* glabrous; anthers almost twice as long as the filaments. *Style* 15−18 mm long. *Chromosome no.* 2n=40.

Found only once on the Bokkeveld Mountains plateau near Nieuwoudtville (3119−CB).

Only voucher: *Burger* STE in 30207.

Readily distinguished by its carmine-red flowers without any blotches or markings inside, its glabrous filaments, and by its corm which has a basal ridge which is not completely circular.

34. **Romulea setifolia** *N.E. Br.* in Gdnrs' Chron. 92: 467 (1932); De Vos in Jl S. Afr. Bot. Suppl. 9: 171, fig. 51 (1972). Type: Cape, Mossel Bay, Gouritz River, *Muir* 4847 (K, holo.!; PRE!).

Plants 50−250 mm long. *Corm* with a small crescent-shaped basal ridge sometimes forming an almost complete circle; tunics minutely pitted, with rows of minute fibre clusters on the ridge. *Stem* short, hidden, or up to 120 mm long. *Basal leaves* 2 or more, filiform, 50−250 × 0,5−1,5 mm, grooves narrow. *Outer bract* largely green; inner with colourless or brown-edged membranous margins, sometimes slightly shorter than the outer. *Flowers* 15−45 mm long, yellow or apricot-yellow, sometimes with dark blotches in the throat. *Perianth tube* 3−7 mm long; segments 8−35 × 2,5−10 mm. *Filaments* 4−8 mm; anthers 2−8 mm long. *Style* 8−14 mm; stigmas more or less at the anther tips. *Capsules* on erect or suberect peduncles. *Chromosome no.* 2n=30, c. 32, c. 54.

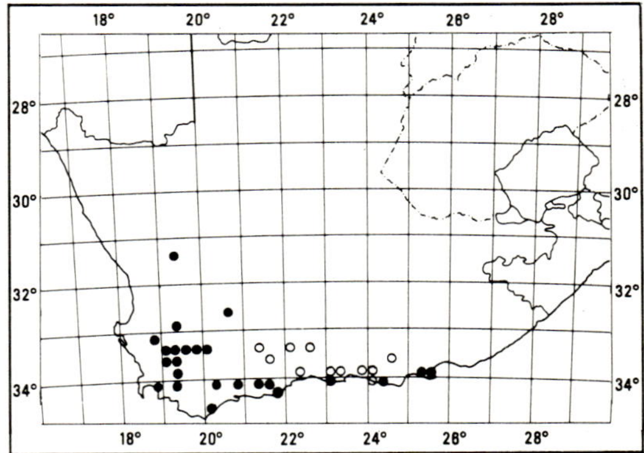

MAP 11.—● **Romulea setifolia**
○ **R. fibrosa**

Found in southern Cape districts from Caledon to Port Elizabeth, and in south-western inland districts from Ceres and Worcester to Laingsburg, with outliers to Calvinia. Map 11.

Four varieties are recognized:

1 Perianth without dark blotches in throat:

  2 Flowers 12−25 mm long; anthers shorter than filaments:

    3 Seeds not sticky; flowers usually yellow, sometimes apricot-yellow ..... (a) var. *setifolia*

FIG. 9.−1, **Romulea amoena**, habit, × 1; 1a, base of corm; 1b, outer perianth segment, upper face (left) and lower face (right); 1c, outer (left) and inner (right) bract (*De Vos* 1601). 2, **R. longipes**, habit, × 7/8; 2a, outer perianth segment, lower face; 2b, outer (left) and inner (right) bract; 2c, ripening capsules; 2d, transverse section of leaf (*De Vos* 2201).

3 Seeds sticky when dry; flowers apricot-
   yellow .... (b) var. *belviderica*

2 Flowers 25 mm or more in length; anthers
   subequal to filaments or slightly
   longer .... (c) var. *aggregata*

1 Perianth with dark blotches in throat
   .... (d) var. *ceresiana*

(a) var. **setifolia.**

Stem short, or sometimes up to 50 mm. *Leaves* less than 1 mm in diam. *Bracts* reaching more than halfway up the perianth segments or almost to their tips. *Flowers* 12−25 mm long, pale yellow or sometimes apricot-yellow. *Perianth segments* 2,5 to almost 6 mm wide *Stamens* 6−10 mm; anthers shorter than the filaments.

Found in south-western inland Cape districts of Ceres and Sutherland and southern districts from Caledon to Port Elizabeth.

Vouchers: *Muir* 996 (BOL; PRE); *Muir* 4847 (PRE; K); *Acocks* 21349 (NBG; PRE); *De Vos* 1931 (STE).

(b) var. **belviderica** *De Vos* in Jl S. Afr. Bot. Suppl. 9: 173 (1972). Type: Cape, Knysna, Belvidere, *Duthie* 1246 (STE, holo.!),

Very closely related to var. *setifolia*, differing mainly in its glutinous seeds, also in its usually apricot-yellow flowers with the perianth segments subobtuse, 5−6 mm wide.

Known only from the Belvidere Estate, west of Knysna.

Vouchers: *Duthie* 1246 (STE); *De Vos* 2073 (STE; PRE).

As it was found impossible to cross this variety with var. *setifolia*, it has been placed in a distinct variety.

(c) var. **aggregata** *De Vos* in Jl S. Afr. Bot. Suppl. 9: 175 (1972). Type: Cape, Worcester, near Breë River bridge NE of Bainskloof, *De Vos* 1276 (STE, holo.!).

Stem short or sometimes up to 120 mm long. *Leaves* 1 (−2) mm diam. *Bracts* seldom reaching more than halfway up the perianth. *Flowers* 25−45 mm long, golden-yellow, sometimes pale apricot, cup bright yellow, sometimes with dark or apricot-coloured veins in the throat. *Perianth segments* (5−) 6−10 mm wide. *Stamens* 10−15 mm long; anthers subequal to or slightly longer than the filaments.

Found in the inland south-western Cape districts of Ceres, Tulbagh and Worcester, also Swellendam, Heidelberg and Calvinia.

Vouchers: *Esterhuysen* 18688, 9249 (BOL); *Marloth* 9084 (STE; PRE); *De Vos* 1885 (STE).

(d) var. **ceresiana** *De Vos* in Jl S. Afr. Bot. Suppl. 9: 175, fig. 52 (1972). Type: Cape, Ceres, flats W of Theronsberg Pass, *De Vos* 1676 (STE, holo.!).

Very closely related to var. *aggregata* from which it differs as follows: *Stem* short, not extending from leaf bases. *Flowers* with a large or small dark median blotch on each segment in the throat from which a slender dark line runs down into the orange-yellow cup. *Anthers* and *stigmas* not reaching halfway up the perianth. *Chromosome no.* 2n=30.

Found in the districts of Ceres, Worcester and Laingsburg.

Vouchers: *Oliver* 3475 (STE); *De Vos* 1571 (STE); *Mauve & Oliver* 184 (PRE; STE).

This variety has a flower resembling that of *R. montana* (no. 11) but it differs in its corm with its fibril clusters, in its inner bracts with colourless margins, as well as in chromosome number.

35. **Romulea longipes** *Schltr.* in J. Bot., Lond. 36: 377 (1898); Bég. in Malpighia 23: 77 (1909); Martin & Noel in Publ. Dept. Bot. Rhodes Univ. 30 (1960); Batten & Bokelmann, Wild Flow. E. Cape. Prov. 32, Pl. 26.3 (1966); De Vos in Jl S. Afr. Bot. Suppl. 9: 181, fig. 54 (1972). Type: Cape, Bathurst, Port Alfred, *Galpin* 3023 (B, holo.!; PRE!); topotype: *Galpin* s.n. anno 1895 (GRA!).

Plants 150−500 mm long. *Corm* with minute fibril clusters on a crescent-shaped basal ridge. *Stem* 40−350 mm long. *Basal leaves* 2, filiform, glabrous or ciliate along the median rib margins, 150−500 × 0,8−1 mm, grooves usually narrow. *Bracts* sub-membranous in the lower half, inner with wide brown-edged or streaked membranous margins. *Flowers* 22−38 mm, long, cream, pale yellow, greenish yellow or pale apricot, with a yellow cup, outer perianth segments often with greenish brown feathered veining on the backs. *Perianth tube* 3−5 mm long; segments 15−30 × 6−8 mm. *Filaments* 3−8 mm; anthers 5−7 mm, at first joined at their tips, yellow or rarely violet. *Style* 8−14 mm; stigmas reaching the anther tips. *Capsules* on straight or slightly patent peduncles. *Chromosome no.* 2n=c. 30. Fig. 9: 2.

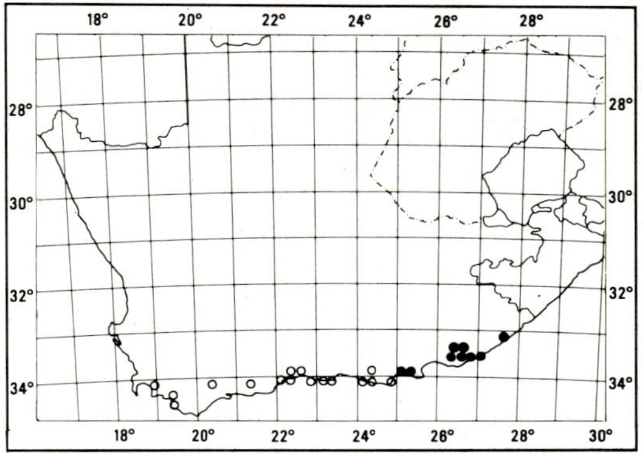

MAP 12.—● **Romulea longipes**
          ○ **R. dichotoma**

Found in eastern Cape coastal districts from Port Elizabeth to East London, in sandy soil at low altitude. Map 12.

Vouchers: *MacOwan* 316 (BOL); *Long* 186 (BOL; GRA; PRE; K), 829 (GRA; K; PRE); *Acocks* 21060; *Jacot Guillarmod* 6692 (GRA; STE); *De Vos* 2201 (STE; PRE).

About 40 km west of Port Elizabeth this species comes into contact with *R. dichotoma* (no. 38) to which it is closely allied. It differs from the latter in its two basal leaves with the leaf ribs almost equal in width, each rib possessing one large and two small veins, in flower colour and in the less marked dichotomous branching of the stem.

36. **Romulea fibrosa** *De Vos* in Jl S. Afr. Bot. Suppl. 9: 183, fig. 55 (1972). Syntypes: Cape, Knysna, on crest of Hoogeberg, *Keet* 1052 (BOL!; GRA!); Uniondale, Bloubosberg, E ridge, *Fourcade* 2831 (BOL, lecto.!; K!).

R. alpina L. Bol. in S. Afr. Gdng 18: 342 (1928), non Rendle (1894). Type as for *R. fibrosa.*

Very closely related to *R. longipes* (no. 35) from which it differs as follows: *Older corm* covered with thick layers of fibrous remains of old leaf bases and with apical fibres 20—80 mm long around the base of the shoot; (younger corm as in *R. longipes). Basal leaves* sometimes more than 2 or only one. Bracts submembranous or greenish, outer with narrow, and inner with wider membranous margins rusty-red in the upper half or sometimes colourless. *Flowers*

magenta to pale pink, with a diffuse violet-blue blotch on each segment in the throat, cup often yellow with small orange markings; outer segments on the backs reddish purple or with 3 violet veins and fine feathered veining towards the margins. *Stigmas* reaching below or almost to the anther tips.

Found at high altitude on southern Cape mountain ranges from George to Humansdorp and Uitenhage, and from Ladismith to Uniondale. Map 11.

Vouchers: *Hutchinson* 1220 (PRE; K); *Esterhuysen* 16406 (BOL; STE), 6523 (BOL); *De Vos* 2078 (STE; PRE); *Stokoe* in SAM 55759 (SAM); *Thompson* 1602 (STE).

Specimens without corms or with young corms can only be distinguished from *R. longipes* (no. 35) by their flower colouring and their more membranous bracts. *R. fibrosa* differs from *R. dichotoma* (no. 38) in usually having 2 basal leaves with the four ribs subequal in width.

37. **Romulea jugicola** *De Vos* in Jl S. Afr. Bot. Suppl. 9: 185, figs 56 & 63 (1972). Type: Cape, George district, 31 km SE of Dysseldorp, *Acocks* 20592 (PRE; holo.!; K!; M!).

Closely related to *R. fibrosa* (no. 36) from which it differs as follows: *Corm* with apical fibres 10—15 mm long. *Stem* angled, often sparsely ciliate on the angles. *Basal leaf* single, 1—2 mm in diam., with white hairs on the margins of the two median ribs, grooves somewhat widened. *Bracts* green, inner with wide, brown-speckled or brown-streaked membranous margins narrowing to a small scarious tip. *Flowers* orange-yellow, outer segments on the backs reddish brown or greenish brown. *Perianth segments* 8—12 mm wide. *Filaments* 6—7 mm, pilose at the bases or up to the tips.

Found on stony foothills of the Kammanassie Mountain in the Klein Karoo between Oudtshoorn and Uniondale (3322—DA).

Vouchers: *Acocks* 20592 (PRE; K; M); *De Vos* 2212 (STE).

Readily distinguished by its corm, single basal leaf with a greater degree of hairiness than in other species, and orange-coloured flowers.

38. **Romulea dichotoma** *(Thunb.) Bak.* in J. Linn. Soc., Bot. 16: 89 (1877); Klatt, in Abh. Naturforsch. Ges. Halle 15: 401 (1882); in Dur. & Schinz, Consp. Fl. Afr. 5: 164 (1895), excl. syn. *R. fragrans;* Bég. in Malpighia 23: 91 (1909), excl. syn. *R.*

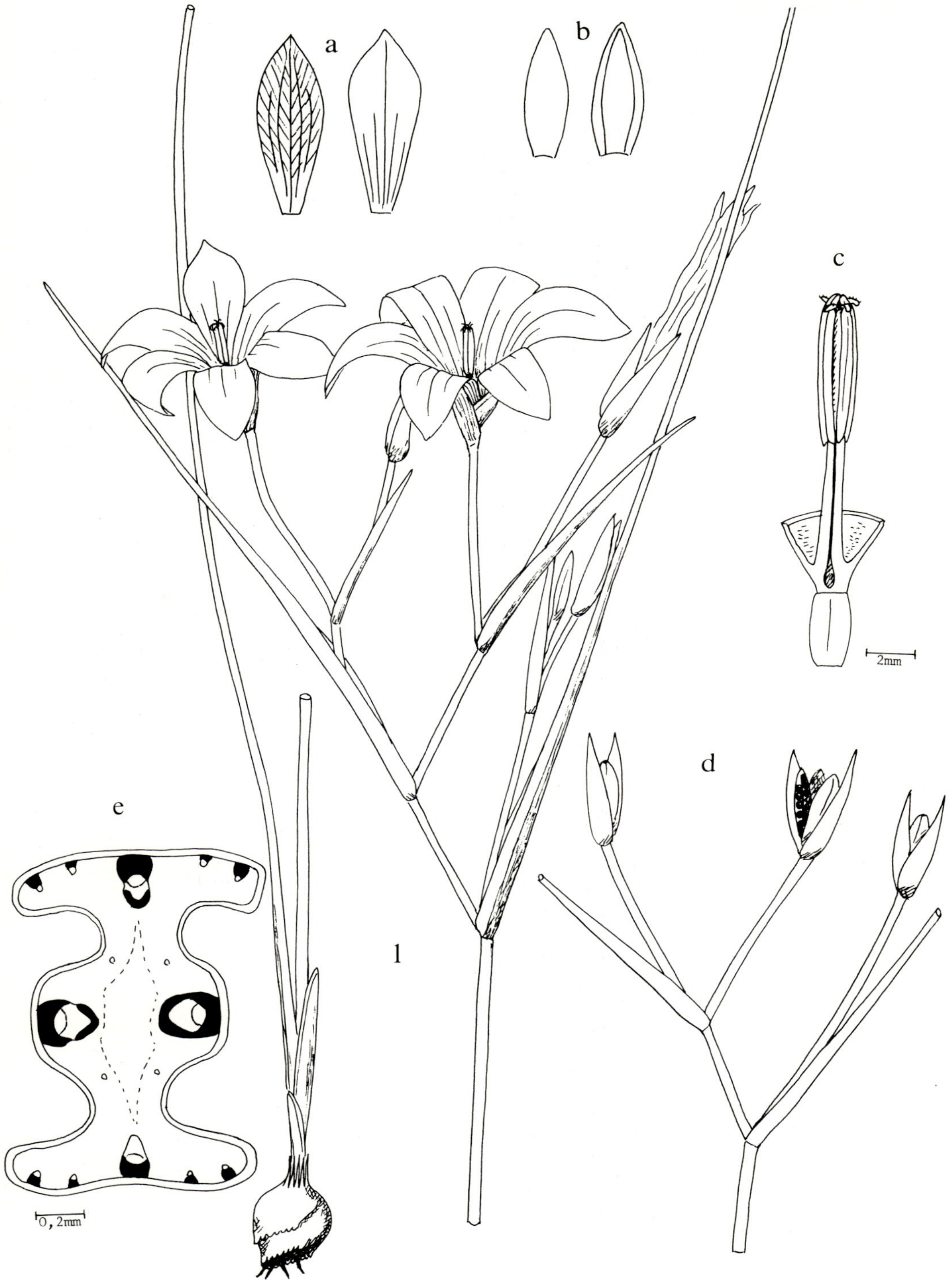

a
b
c
2mm
d
e
l
O,2mm

fragrans; N. E. Br. in J. Linn. Soc., Bot. 48: 21 (1928); De Vos in Jl S. Afr. Bot. Suppl. 9: 187, figs 57 & 61 (1972). Type: Cape, *Thunberg* s.n. herb. no. 1019 (UPS, holo.!).

Gladiolus dichotomus Thunb., Diss. Glad. 10 (1784); Fl. Cap. 1: 187 (1811); Spreng., Syst. Veg. edn 16,1: 152 (1825).

Trichonema dichotomum Klatt in Linnaea 34: 666 (1865−66), excl. syn. *T. fragrans* Eckl. *Bulbocodium dichotomum* (Klatt) Kuntze, Rev. Gen. 2: 700 (1891). *Romulea rosea* var. *dichotoma* (Klatt) Bak., Handb. Irid. 104 (1892), excl. syn. *R reflexa* & *R. tubata*. Type: Cape, Clanwilliam, Olifantrivier & Villa Brakfontein (locality probably incorrect), *Ecklon & Zeyher* s.n. (C, holo.!).

R. caplandica Bég. in Bot. Jb. 38: 332 (1907); in Malpighia 23: 112 (1909). Type: Cape, Riversdale, *Rust* 566B (B, holo.!).

Closely related to *R. longipes* (no. 35) from which it differs as follows: *Stem* usually dichotomously branched near the top, the branches more or less divaricate. *Basal leaf* single (or rarely two in young plants, the first leaf then shorter than the second), filiform or somewhat 4-angled, 1−2 mm wide, with the two lateral ribs narrower than the two median ones, grooves narrow or wide. *Bracts* green, inner with wide membranous margins which are colourless in the lower half and reddish brown in the upper. *Flowers* pink or rarely salmon-pink, with 1−3 pink-purple veins in the throat, cup greenish or golden-yellow with V-shaped marks, outer segments on the backs mostly with 3−5 reddish brown veins and fine feathered veining. *Capsules* on straight, patent peduncles. *Chromosome no.* 2n=30. Fig. 10.

Widespread in southern Cape coastal districts from Caledon to Humansdorp. Map 12.

Vouchers: *Duthie* 1243 (STE); *Rodin* 1322 (BOL; PRE; K; UC); *Muir* 1595 (BOL; GRA; NU; Z); *Fourcade* 5612 (NBG); *De Vos* 2063 (STE; PRE).

This species has a more western distribution than *R. longipes*. The areas of the two species adjoin about 40 km W of Port Elizabeth.

39. **Romulea albomarginata** *De Vos* in Jl S. Afr. Bot. Suppl. 9: 177, fig. 53 (1972). Type: Cape, Ceres Koue Bokkeveld, 37 km N of Gydouw Pass, *De Vos* 1999 (STE, holo.!).

Plants 120−250 mm long. *Corm* with a crescent-shaped basal ridge with rows of minute fibril clusters. *Stem* short or sometimes up to 50 mm. *Basal leaves* 2 or more, filiform, 120−250 × 0,8−1 mm, grooves narrow. *Bracts* largely green or purplish, inner with wide, colourless membranous margins and green tip. *Flowers* 20−38 mm long, bright magenta-pink with dark veins in the throat and orange-yellow cup, outer perianth segments maroon on the backs. *Perianth tube* 4−6 mm long; segments 15−25 × 5−9 mm. *Filaments* 3−4 mm, orange-yellow, pilose in the lower half; anthers 4−7 mm long, pale yellow. *Style* 10−12 mm; stigmas more or less at anther tips. *Capsules* ellipsoid, on widely patent peduncles which straighten later. *Chromosome no.* 2n=30.

Found only on the plateau of the Ceres Koue Bokkeveld (3219−CC, CA).

Vouchers: *Schlechter* 8910 (BOL; GRA); *De Vos* 1956, 2056 (STE); *Hanekom* 618.

The flower somewhat resembles that of *R. fibrosa* (no. 36) and of *R. saxatilis* (no. 3). The species is distinguished in its generally shorter stem and in leaf structure, and from the former also in its smooth, hard corm tunics and free anthers; from the latter in its corm tunics with its fibril clusters on the basal ridge and shorter apical fibres, and in its filaments being pilose only in the lower half.

3. Section **Tortuosae**

Tortuosae *Bég.* in Malpighia 23: 96 (1909); De Vos in Jl S. Afr. Bot. Suppl. 9: 151 (1972). Type species: *R. tortuosa* (Licht. ex Roem. & Schult.) Bak.

Corm laterally or obliquely flattened, often almost lens-shaped, with a wide, somewhat vertical, often fan-shaped ridge across the base and up the sides; tunics split on the ridge

---

FIG. 10.−1, **Romulea dichotoma**, habit, × 1; a, outer (left) and inner (right) perianth segments, lower face; b, outer (left) and inner (right) bract; c, perianth tube, stamens and pistil; d, capsules; e, transverse section of leaf (*De Vos* 2063).

1b
2b
2c
1c
1
1a
1d
0,2mm
2
2a
2d
2mm

into fine parallel fibrils often clustered into irregular groups. *Stem* short, hidden by leaf bases. *Foliage leaves* basal, often spirally twisted or flexuose, or sometimes curved or suberect, largely bifacial and conduplicate with an adaxial groove, and with 4 stomatiferous grooves. *Bracts* membranous or submembranous in the lower half, green or greenish towards the acute or acuminate tip. *Flowers* yellow. *Perianth tube* short, or rarely as long as the segments or longer. *Capsules* ellipsoid to globose, on recurved or spirally twisted peduncles. *Chromosome no.* 2n=30, 32.

Four species, one of which (*R. macowanii*, no. 43) has previously been placed in a separate subsection on account of its long perianth tube. Found on Cape inland plateaux and mountain ranges mostly at altitudes of 800−1700 m altitude; also in Lesotho.

40. **Romulea austinii** *Phill.* in Flower. Pl. Afr. 3: t. 90 (1932); De Vos in Jl S. Afr. Bot. Suppl. 9: 153, fig. 44 (1972). Type : Cape, Laingsburg, Matjiesfontein, *Austin* 2572 (PRE, holo.!; BOL!; K!).

Plants 60−120 mm long. *Corm* obliquely flattened in the lower half with a wide crescent-shaped basal ridge. *Leaves* filiform, suberect, bent or slightly flexuose, 40−200 × 0,5−1 mm, sometimes ciliolate on the rib margins, grooves narrow. *Bracts* green in the upper half, submembranous in the lower, inner with wide brown-edged or speckled membranous margins. *Flowers* 20−33 mm long, yellow, usually with a spade-shaped brownish black blotch on each perianth segment, outer segments dark-veined or greenish brown on the backs. *Perianth tube* 4−5 mm long; segments 14−25 × 5−9 mm. *Filaments* 5−7 mm, slightly widened at the bases; anthers 3−6 mm long. *Style* 9−14 mm; stigmas more or less at the anther tips. *Capsules* ellipsoidal, on curved, slightly flexuose peduncles. *Chromosome no.* 2n=30. Fig. 11:1.

Found in the western, south-western and southern parts of the Great Karoo from Calvinia to Laingsburg and to Uniondale, also near Montagu and Port Elizabeth. Map 13.

Vouchers: *Taylor* 446 (BOL); *Salter* 1053 (BOL); *Compton* 10857 (NBG); *Hall* 2379 (NBG); *De Vos* 1950 (STE).

This species links section *Tortuosae* and section *Romulea*, as its corm is intermediate in shape, with a ridge not as widely expanded as in *R. tortuosa* (no. 41), and as its bracts are greener. The yellow colour of the flowers tends to fade in herbarium specimens exposed to light.

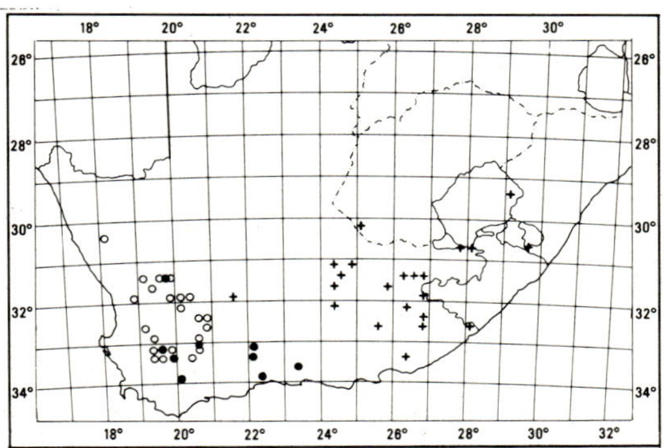

MAP 13.—● **Romulea austinii**
      ○ **R. tortuosa**
      + **R. macowanii**

41. **Romulea tortuosa** *(Licht. ex Roem. & Schult.) Bak.* in J. Linn. Soc., Bot. 16: 88 (1877); Handb. Irid. 100 (1892); in F. C. 6: 37 (1896); Klatt in Abh. Naturforsch. Ges. Halle 15: 399 (1882); in Dur. & Schinz, Consp. Fl. Afr. 5: 167 (1895); Bég. in Malpighia 23: 96 (1909); De Vos in Ann. Univ. Stell. 28A, 3: 77, fig. 7 (1952); in Jl S. Afr. Bot. Suppl. 9: 154, fig. 45−47 (1972). Type: Cape, C. B. S., sub *Ixia tortuosa* (B, lecto.!).

Plants 50−250 mm long. *Corm* laterally flattened, almost lens-shaped, with a wide fan-shaped ridge. *Leaves* filiform, spirally twisted, flexuose or sometimes suberect, 80−250 × 0,5−1 mm, sometimes ciliolate on the rib margins, grooves narrow. *Bracts* membranous or submembranous, greenish or with green or reddish veins towards the tip. *Flowers* 15−55 mm long, pale to bright yellow, often with a spade-shaped or tridentate black blotch or with dark veins on each segment. *Perianth tube* 3−10 mm long;

FIG. 11.−1, **Romulea austinii**, habit, × 1; 1a, corm seen from the opposite side; 1b, outer (left) and inner (right) bract; 1c, mature capsules; 1d, transverse section of leaf (*De Vos* 1950) 2, **R. tortuosa** subsp. **tortuosa**, habit, × 1; 2a, corm seen from the side; 2b, mature capsules; 2c, bracts; 2d, perianth tube, stamens and pistil (*De Vos* 1948).

segments 10—40 × 2—12 mm. *Filaments* 3—10 mm, often slightly widened at the bases; anthers 3—10 mm long. *Style* 7—20 mm; stigmas at the anther tips or higher. *Capsules* subglobose to ellipsoidal, on reflexed or spirally twisted peduncles. *Chromosome no.* 2n=30.

Found in the inland districts of the south-western Cape, and western Karoo. Map 13.

Three subspecies are recognized, differing mainly in flower size and in the shape and markings on the perianth:

1 Flowers usually more than 25 mm long, bright golden-yellow, with segments more than 5 mm wide:

  2 Perianth generally with dark blotches or veins in the throat; segments widest just above the middle ....(a) subsp. *tortuosa*

  2 Perianth without dark blotches or veins in the throat; segments widest in the upper quarter................................(b) subsp. *aurea*

1 Flowers 15—20 (—25) mm long, usually pale yellow, with segments 2—4 mm wide .... (c) subsp. *depauperata*

(a) subsp. **tortuosa.**

Ixia tortuosa Licht. ex Roem. & Schult., Syst. Veg. 1: 375 (1817); Licht., Reisen im südlichen Afrika 2: 289 (1812), nom. nud. *Trichonema tortuosum* Ker-Gawl., Irid. Gen. 83 (1827); Klatt in Linnaea 34: 666 (1865-66). *Bulbocodium tortuosum* (Bak.) Kuntze, Rev. Gen. Pl. 2: 701 (1891).

R. tridentifera Klatt in Abh. Naturforsch. Ges. Halle 15: 398 (1882); in Dur. & Schinz, Consp. Fl. Afr. 5: 167 (1895); Bak., Handb. Irid. 101 (1892); in F. C. 6: 39 (1896); Bég. in Malpighia 23: 97 (1909). Type: Cape, Calvinia, Hantam Mtns, *Meyer* s.n. anno 1869 (B, holo.†; S!).

Flowers 25—37 mm long, rarely smaller, bright golden-yellow, the cup orange-yellow. *Perianth segments* acute to subacute, rarely subobtuse, 6—10 mm wide, widest just above the middle, with a black blotch or 1—3 black veins on each segment. *Style* 10—16 mm long. Fig. 11: 2.

Found from Calvinia to Sutherland and Laingsburg.

Vouchers: *Salter* 6681 (BOL); *Barker* 10771 (NBG); *De Vos* 1572 (STE); *Acocks* 18211.

(b) subsp. **aurea** *(Klatt) De Vos* in Jl S. Afr. Bot. Suppl. 9: 157, figs 34 & 46 (1972).

Type: Cape, Calvinia, Hantam Mtns, *Meyer* s.n. anno 1869 (B, holo.!).

R. aurea Klatt in Abh. Naturforsch. Ges. Halle 15: 399 (1882); in Dur. & Schinz, Consp. Fl. Afr. 5: 162 (1895).

Flowers 30—55 mm long, fragrant, buttercup-yellow to almost cadmium-orange, without dark marks, the upper part of the segments paler yellow. *Perianth segments* 7—12 mm wide, widest in the upper quarter or third, obtuse, frequently apiculate. *Style* 14—20 mm long.

Found in the Calvinia district.

Vouchers: *Salter* 3501 (BOL; K); *Schlechter* 10894 (BOL; GRA; PRE; G; K; S; etc.); *Marloth* 10251 (PRE; STE); *De Vos* 1600 (STE).

(c) subsp. **depauperata** *De Vos* in Jl. Afr. Bot. Suppl. 9: 159, fig. 47 (1972). Type: Ceres, summit of Gydouw Pass, *De Vos* 1273 (STE, holo.!).

Flowers 15—20 (—25) mm long, pale yellow, with a small dark blotch or 1—3 dark veins halfway up each perianth segment. *Perianth segments* 2—4 mm wide, acute to acuminate. *Style* 7—10 mm long.

Found in the Ceres, Worcester and Laingsburg districts.

Vouchers: *Esterhuysen* 17381 (BOL); *Salter* 2626 (BOL; K); *De Vos* 1797 (STE); *Van Breda & Joubert* 1887.

42. Romulea sphaerocarpa *De Vos* in Jl. S. Afr. Bot. Suppl. 9: 161, figs 35 & 48 (1972). Type: Cape, Worcester, Sandvlei, 23 km from Matroosberg Station on road to Koo, *De Vos* 2102 (STE, holo.!).

Plants 150—300 mm long. *Corm* laterally flattened, almost lens-shaped, with a wide fan-shaped ridge. *Stem* elongating up to 40 mm after flowering. *Leaf* single, rarely 2, filiform, bent or suberect, 150—300 × c. 1 mm, glabrous, with adhering sand grains, grooves narrow. *Inner bract* purplish brown or green in upper half, with wide, usually colourless membranous margins. *Flower* usually single, 25—35 mm long, golden-yellow with brownish veins in the throat, outer perianth segments with 5 brown veins and feathered veining on the backs.

FIG. 12.—1, **Romulea macowanii** var. **macowanii**, habit, × c.1; a, part of perianth; b, perianth tube, stamens and pistil; c, bracts; d, mature capsules; e, transverse section of leaf (*De Vos* 2258).

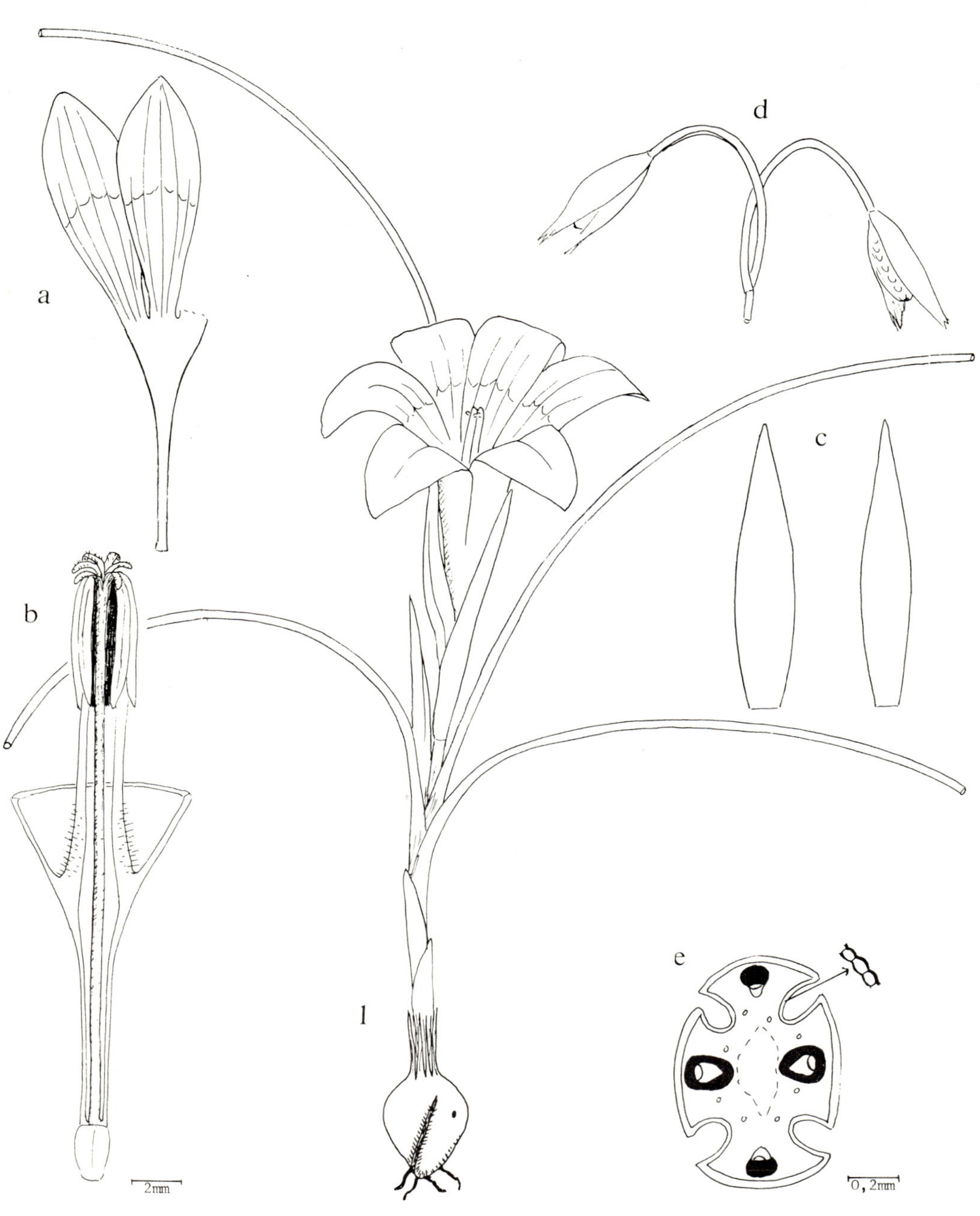

a
b
d
c
l
e
2mm
0,2mm

Perianth tube 5—6 mm long; *segments* 15—25 × 5—12 mm. *Filaments* 5—7 mm; anthers 6—8 mm. *Style* 10—14 mm; stigmas below to above anther tips. *Capsule* spherical, on an arcuate peduncle. *Chromosome no.* 2n=c. 30, 32.

Found in the south-western part of the Great Karoo, near and south of Matroosberg Station. (3319—BD, DB).

Vouchers: *Stayner* s.n., June 1970 (NBG); *De Vos* 2102, 2191 (STE).

An early flowering species (June), locally frequent on stony ridges in sand pockets, and distinguished by its usually single yellow flower and single leaf. The leaf and bracts exude some mucilage.

43 **Romulea macowanii** *Bak.* in J. Bot., Lond. 5: 236 (1876); in J. Linn. Soc., Bot. 16: 88 (1877); Handb., Irid. 101 (1982); in F.C. 6: 38 (1896); Klatt in Dur. & Schinz, Consp. Fl. Afr. 5: 165 (1895); Bég. in Malpighia 23: 113 (1909); B. L. Burtt in Curtis's bot. Mag. 176: t. 515 (1967); De Vos in Jl S. Afr. Bot. Suppl. 9: 165, fig. 49 (1972). Type: Cape, Somerset East, in graminosis summi Boschberg, *MacOwan* 1547 (K, lecto.!; BOL!; GRA!; PRE!; BM!; G! etc.).

Plants 200—400 mm long. *Corm* flattened on one side, with a small fan-shaped ridge. *Leaves* filiform, curved or suberect, 150—400 × 0,5—1 mm, grooves narrow. *Inner bract* with a green tip and sometimes with wide, white membranous margins. *Flowers* 22—105 mm long, golden-yellow, often orange-yellow in the lower half, outer perianth segments brownish or purplish on the backs or with reddish feathered veining. *Perianth tube* (13—) 20—65 mm long, tubular for most of its length, widened and funnel-shaped in the upper 10 mm; segments (10—) 15—45 × 5—15 mm. *Filaments* 5—10 mm, inserted in the upper part of the perianth tube; anthers 5—12 mm long. *Style* (20—) 25—70 mm; stigmas below, at or sometimes above the anther tips. *Capsules* ellipsoid, on recurved peduncles.

A montane species widely distributed on inland mountains at high altitude in the eastern Karoo, midlands and north-eastern part of the Cape Province and in Lesotho, with outliers to Fraserburg and Philippolis. Map 13.

Three varieties are recognized:

1 Perianth tube 13—33 mm long, about as long as the segments or 1,5 times their length; bracts reaching almost to the bases of the segments or higher:

  2 Flowers (45—) 55—80 mm long; perianth segments 8—15 mm wide ..... (a) var. *macowanii*

  2 Flowers 27—55 (—60) mm long; perianth segments 4—7 mm wide .... (c) var. *oreophila*

1 Perianth tube 35—65 mm long, about twice as long as the segments or rarely 1,7 times their length; bracts reaching three-quarters to halfway up the perianth tube ..... (b) var. *alticola*

(a) var. **macowanii**.

Bulbocodium macowanii (Bak.) Kuntze, Rev. Gen. Pl. 2: 700 (1891).

Syringodea luteo-nigra Bak. in Kew Bull. 1897: 281 (1897). Type: Cape, Queenstown, summit of Andriesberg and of Hangklip, *Galpin* 1516 (K, holo.!; PRE!; BOL!).

Bracts green in upper half or three-quarters, submembranous in lower part, reaching more or less to the bases of the perianth segments; inner bract often with wide, white membranous margins. *Flowers* (45—) 55—80 mm long. *Perianth tube* 13—33 mm long, equal to or slightly shorter than the segments; segments 25—45 × 8—15 mm. *Style* 30—50 mm long. Fig. 12.

Found on high inland plateaux, generally at altitudes of 1 500—2 000 m, from Fraserburg, Philippolis and Lesotho to Cathcart, and also near Grahamstown.

Vouchers: *Bolus* 592 (BOL); *De Vos* 2258 (STE); *Galpin* 1516 (BOL; PRE; K); *Acocks* 14294.

(b) var. **alticola** (*B. L. Burtt*) *De Vos* in Jl S. Afr. Bot. Suppl. 9: 168 (1972). Type: Lesotho, *Milford* s.n., cult. *Stern* 9—10—1963 (K, holo.!).

R. longituba L. Bol. var. *alticola* B. L. Burtt in Curtis's bot. Mag. 176: t. 515 (1967); Hilliard & B. L. Burtt in Notes R. bot. Gdn Edinb. 30: 126 (1970).

R. longituba L. Bol. in S. Afr. Gdng 18: 341 (1928). *R. longituba* G. J. Lewis in Jl S. Afr. Bot. 7: 43 (1941); l.c. 14: 89 (1948). Type: Cape, Griqualand East, Mt Ingeli, *Tyson* 1267 (SAM, holo.!; GRA!).

Bracts herbaceous or membranous in lower half, green towards the tip, reaching halfway to three-quarters up the perianth tube. *Flowers* 55—105 mm long. *Perianth tube* 35—65 mm long, about twice as long as the segments; segments 15—35 × 4—10 mm. *Style* 35—70 mm.

From the high mountains of Griqualand East and Lesotho.

Vouchers: *Tyson* 1267 (SAM; GRA); *Schurr* 5 (STE; UN); *Milford* s.n. (K).

(c) var. **oreophila** *De Vos* in Jl S. Afr. Bot. Suppl. 9: 169 (1972). Type: Cape, Barkley East-Maclear boundary, Naudesnek, summit of pass, *De Vos* 2186 (STE, holo.!).

Bracts white and membranous in lower half, green towards the tip, reaching more or less the bases of the perianth segments. *Flowers* 27−55 (−60) mm long. *Perianth tube* 13−27 mm long; segments 10−25 (−30) × 4−7 mm. *Style* 18−30 mm long.

From high mountains of north-eastern Cape Province in districts of Molteno to Maclear and Lesotho.

Vouchers: *Galpin* 6848 (BOL; SAM; PRE; K); *Marais* 1367 (PRE; K); *Jacot Guillarmod* 978.

An alpine variety mostly at altitudes above 2 400 m, with flowers smaller than in var. *alticola*.

4. Section **Hirtae**

Hirtae (*Bég.*) *De Vos* in Jl S. Afr. Bot. Suppl. 9: 269 (1972). Type species: *R. hirta* Schltr.

'Stirps' *Hirtae* Bég. in Ann. Conserv. Jard. bot. Genève 11−12: 159 (1908).

Corm with bent basal teeth over a rounded base or with minute parallel fibrils on a crescent-shaped basal ridge. *Stem* short, hidden. *Foliage leaves* basal, with lateral ribs reduced and medium ribs widened to form 4 longitudinal wings, usually ciliate on the wing margins, up to 5(−7) mm wide. *Flowers* yellow or pink. *Perianth* tube short.

The two species of the section are greatly dissimilar in their corms and flowers; but they correspond so closely in leaf structure, both morphologically and anatomically, as well as in chromosome number, that they have been placed in one section. They are apparently not closely related to other species of *Romulea*.

Found in inland western Cape districts from Calvinia south-eastwards to Swellendam, excluding the extreme SW Cape.

44. Romulea hirta *Schltr.* in Bot. Jb. 27: 91 (1900); Bég. in Malpighia 23: 82 (1909); De Vos in Ann. Univ. Stell. 28A, 3: 80 (1952); in Flower. Pl. Afr. 29: t. 1137 (1952); in Jl S. Afr. Bot. Suppl. 9: 269 (1972). Type: Cape, Clanwilliam, Koudeberg, *Schlechter* 8766 (B, holo.!; BM!; K!; P!; PRE!; S!; Z!).

Plants 100−300 mm long. *Corm* subglobose or ovoid, with curved basal teeth bent over a rounded base. *Stem* short, hidden. *Leaves* 4-winged, ciliate on the wing margins or glabrous, 100−300 × 2−5 mm. *Peduncles* 40−200 mm long. *Bracts* greenish, sometimes submembranous towards the base, inner with wide membranous margins brownish towards the scarious tip. *Flowers* 18−35 mm long, pale yellow, often with a pale reddish brown or greenish brown transverse zone above the throat. *Perianth tube* 4−5 mm long; segments 12−25 × 4−8 mm. *Filaments* 5−6 mm; anthers 3−5 mm long, golden-yellow. *Style* 8−14 mm; stigmas reaching halfway up the anthers or almost to their tips. *Capsules* subglobose or ellipsoid, on suberect or slightly bent peduncles. *Chromosome no.* 2n=26.

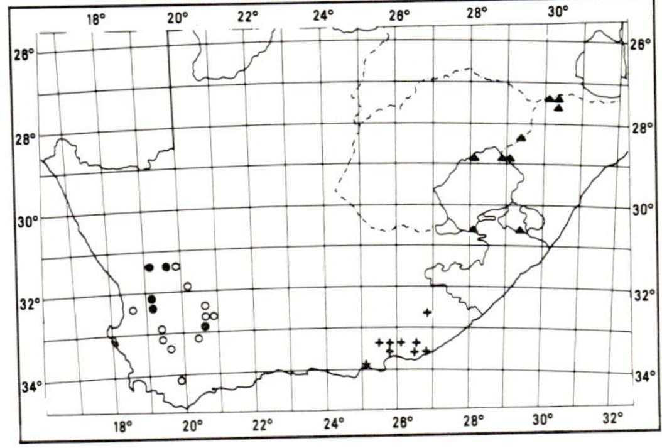

MAP 14.—● **Romulea hirta**
    ○ **R. tetragona**
    + **R. autumnalis**
    ▲ **R. campanuloides**

Found in the western Karoo from Calvinia to Sutherland, also on the Clanwilliam mountain plateaux, on stony ground. Map 14.

Vouchers: *Lewis* 5835 (NBG; STE); *De Vos* 2022 (STE); *Leipoldt* in BOL 20771; *Barker* 10731 (NBG).

Readily distinguished in the live state by its 4-winged leaves and yellow flowers. In herbarium specimens the wings, being pressed flat, are not easily discernible.

45 Romulea tetragona *De Vos* in Flower. Pl. Afr. 29: t. 1136 (1952); in Jl S. Afr. Bot. Suppl. 9: 271, fig. 91 (1972). Type: Cape, Ceres, Theronsberg Pass, *De Vos* 1569 (STE, holo.!).

Plants 80–300 mm long. *Corm* with a crescent-shaped basal ridge, tunics often split into minute parallel fibrils on the ridge. *Stem, leaves* and *peduncles* as in *R. hirta* (no. 44). *Bracts* green, outer with hairs on some veins, with narrow (outer) and wide (inner) membranous margins and tips. *Flowers* 16–35 mm long, violet-rose to lilac or rarely salmon-pink, cup violet or greenish yellow, with a violet blotch or transverse band on each segment in the throat. *Anthers* 2–6 mm long, incurved, circinal or suberect, purple or golden-yellow, 2–6 mm long, pollen brick-coloured, orange or yellow. *Style* 5–12 mm long; stigmas reaching below to just above the anther tips. *Capsules* ellipsoid or subglobose, on curved peduncles which straighten somewhat later. *Chromosome no.* 2n=26.

From the Clanwilliam to Laingsburg districts and to Swellendam. Map 14.

Variation occurs in size and colouring of the perianth and anthers, colour of the pollen, and length of style and style branches. Two varieties are recognized (see descriptions below for distinguishing characters):

(a) var. **tetragona**.

Perianth violet-rose to lilac-pink. *Anthers* purple, incurved or circinal; pollen brick or orange. *Style* branches up to 5 mm long, attenuate.

From Clanwilliam, Ceres and Sutherland.

Vouchers: *Leipoldt* 3823 (BOL); *Lewis* 2632 (SAM); *Hall* 3257 (NBG).

(b) var. **flavandra** *De Vos* in Jl S. Afr. Bot. Suppl. 9: 273 (1972). Type: Cape, Laingsburg, Tweedside, *De Vos* 1800B (STE, holo.!).

Perianth lilac-pink or rarely salmon-pink. *Anthers* suberect or slightly incurved at their tips, golden-yellow, pollen golden-yellow. *Style* branches 2–3 mm long.

From Sutherland, Laingsburg and Swellendam.

Vouchers: *Acocks* 16986 (PRE; K); *De Vos* 1936 (STE); *Neethling* in BOL 24789 (BOL).

5. Section **Roseae**

Roseae *(Bég.) De Vos* in Jl S. Afr. Bot. Suppl. 9: 202 (1972). Type species: *R. rosea* (L.) Eckl.

'Stirps' *Roseae* Bég. in Malpighia 23: 60 (1909), partly.

Corm with a rounded or pointed base, with tunics split at the base into bent or straight acuminate teeth. *Stem* usually short, hidden by the leaf bases. *Foliage leaves* generally basal, terete or compressed cylindrical, or rarely ×-shaped in transverse section, up to 4 mm in diameter, grooves narrow or sometimes wide. *Flowers* variously coloured, often with dark blotches in the throat. *Perianth* tube short. *Capsules* on recurved or sometimes spirally twisted or straight peduncles. *Chromosome no.* 2n=18, 20, 22.

The 16 species of the section have previously been placed in five subsections which are distinguished mainly by the shape of the corm base and the peduncles, the texture of the bracts, and the chromosome number.

Widely distributed throughout the western, south-western to south-eastern Cape Province, mainly from Calvinia to the Cape Peninsula and to Grahamstown; one species also on the East African mountains to the equator.

46. Romulea autumnalis *L. Bol.* in J. Bot., Lond. 69:12 (1931); Martin & Noel in Publ. Dept Bot. Rhodes Univ. 30 (1960); De Vos in Jl S. Afr. Bot. Suppl. 9: 203, fig. 67 (1972). Syntypes: Cape, Grahamstown commonage, *Dyer* 2414 (BOL, lecto.!); 35

FIG. 13.—1, **Romulea autumnalis**, habit, × 7/8; 1a, corm seen from the opposite side; 1b, outer (left) and inner (right) bract; 1c, outer perianth segment, lower face; 1d, perianth tube, stamens and pistil (*De Vos* 1733). 2, **R. campanuloides** var. **campanuloides**, habit, × 1; 2a, outer (left) and inner (right) bract; 2b, outer perianth segment, lower face; 2c, perianth tube, stamens and pistil; 2d, transverse section of leaf (*De Vos* 2187).

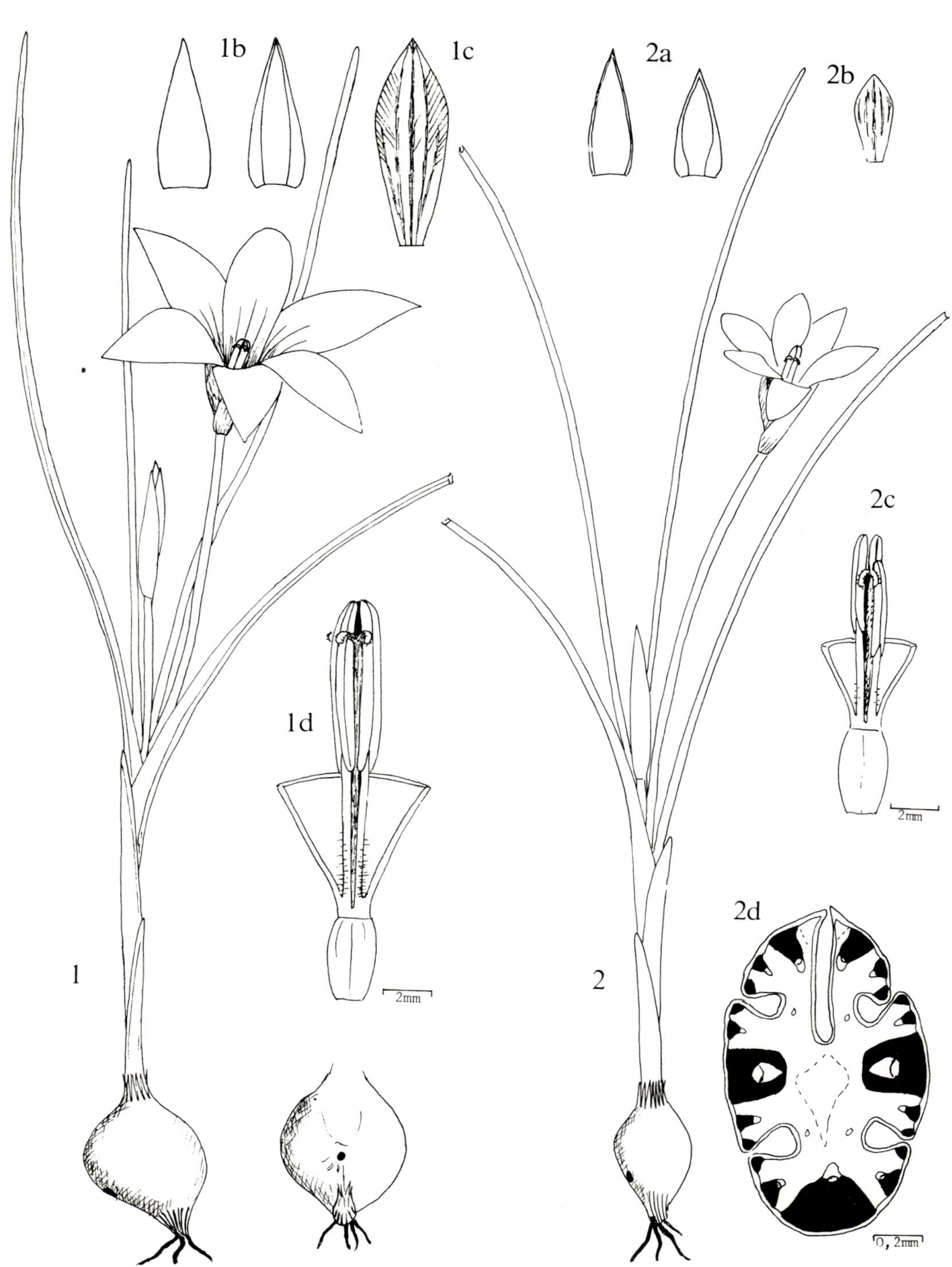

1b
1c
2a
2b
2c
1d
1
2
2d
2mm
2mm
2mm
0,2mm

km from Grahamstown towards Kariga, *Dyer* 2421 (BOL!).

R. rosea Eckl. var. *speciosa* sensu Bak. in F.C. 6: 42 (1896), partly.

Plants 150−350 mm long. *Corm* obovoid, obliquely pointed at the base, with short basal fibrils on a very small basal ridge. *Leaves* basal, filiform to compressed cylindrical, 100−350 × 1−2 mm, grooves narrow. *Peduncles* 50−180 mm long. *Bracts* green or greenish, outer with narrow, inner with wide, colourless membranous margins narrowing to a green tip. *Flowers* 25−40 mm long, pink, magenta-pink or rarely white, with a yellow or orange-yellow cup, outer segments green on the backs with 3−5 purplish veins and feathered veining. *Perianth tube* 5−7 mm long; segments 18−30 × 6−12 mm. *Filaments* 4−10 mm; anthers 6−8 mm long, reaching less than halfway up the perianth. *Style* 8−12 mm; stigmas at or just below the anther tips. *Capsules* on erect peduncles. *Chromosome no.* 2n=22. Fig. 13:1.

In the eastern Cape, from Somerset East and Uitenhage to Bathurst and King William's Town, at low altitudes and up to 650 m, on grassy flats and mountain slopes. Map 14.

Vouchers: *Cruden* 248 (GRA; STE); *Makungo* 78 (STE; Fort Hare); *Dyer* 2421 (BOL); *Cheadle* 718; *Bayliss* 4453 (NBG).

This generally lowland species, as well as the closely related montane *R. campanuloides* (no. 47), differs from *R. rosea* (no. 59), *R. eximea* (no. 61), and *R. cruciata* (no. 60) which have somewhat similar flowers, in the small ridge at the base of the corm, and inner bracts with colourless membranous margins; also in flowering period (late summer and autumn) and in chromosome number.

47. Romulea campanuloides *Harms* in Bot. Jb. 19 Beibl. 47: 28 (1894); Bak. in F.T.A. 7: 345 (1898); Bég. in Malpighia 23: 74 (1909); De Vos in Jl S. Afr. Bot. Suppl. 9: 207, fig. 68 (1972). Type: Kenya, Kilimanjaro, Bergwiese oberhalb des Urwaldes, *Volkens* 782 (B, holo.†; K, lecto.!; BM!).

Closely related to *R. autumnalis* (no. 46), differing as follows: Plants 100−550 mm long. *Leaves* 80−550 × 0,8−2 mm. *Bracts* green, with narrow, colourless, membranous margins. *Flowers* 15−35 mm long, with a yellow or greenish yellow cup, outer segments magenta or greenish on the

backs, sometimes striped with 3−5 purplish lines. *Perianth segments* 9−25 × 4−8 mm. *Anthers* usually reaching more than halfway up the perianth. *Style* to 17 mm long; stigmas below, at, or above the anther tips. Fig. 13.

On stony or grassy plateaux of the Drakensberg in South Africa and Lesotho, and extending northwards on the East African mountain ranges into Zimbabwe, Malawi, Tanzania and Kenya, at 1 500−3 000 m altitude. Map 14.

This is the only South African species of *Romulea* that reaches beyond the borders of the Republic. Two varieties are recognized:

1 Style 7−10 mm long; stigmas rarely above anther tips; flowers mostly less than 25 mm long ....(a) var. *campanuloides*

1 Style 15−17 mm long; stigmas mostly overtopping anthers; flowers 25−35 mm long .... (b) var. *gigantea*

(a) var. **campanuloides.**

R. alpina Rendle in J. Linn. Soc., Bot. 30: 376 & 401 (1895), non L. Bol. (1928). Type: Kenya, Kilimanjaro, higher slopes to 10 000 ft, *Taylor* 1888 (BM, holo.!).

R. thodei Schltr. in J. Bot., Lond. 36: 318 (1898); Trauseld, Wild Flow. Natal Drakensberg 34, 35 (1969). Type: Orange Free State, Mont aux Sources, wet sandy places on summit, *Thode* s.n., Jan. 1896 (B, holo.!; BOL!; STE!).

R. rosea sensu Bak. in F. C. 6: 42 (1896), partly; Wood in Trans. S. Afr. phil. Soc. 18: 232 (1906), non Eckl. (1827).

R. linaresii Parl. subsp. *abyssinica* sensu Norlindh & Weimarck in Bot. Notiser 1937: 173 (1937), non Parl., nec Bég.

Small plants with all the organs small, e.g. *leaves* 0,8−1 mm in diam. *Flowers* 15−28 mm long, usually with a greenish yellow cup. *Perianth segments* 4−5 mm wide. *Stamens* reaching three-quarters up the perianth; anthers 3−5 mm long. *Style* 7−10 mm long. Fig. 13: 2.

From the Barkley East-Maclear mountains in the north-eastern Cape Province to Kilimanjaro in Kenya.

Vouchers: *Hilliard* 5411 (STE; NU); *De Vos* 2187 (STE); Jacot *Guillarmod* 2096A; *Norlindh & Weimarck* 5055 (SAM; BM; LD; S); *Thode* 58 (BOL).

(b) var. **gigantea** *De Vos* in Jl S. Afr. Bot. Suppl. 9: 209 (1972). Type: Natal, Altemooi, alt. 6 500 ft, *Thode* in STE 3924 (STE, holo.!).

R. thodei Schltr. subsp. *gigantea* De Vos in Jl S. Afr. Bot. 21: 106, fig. 3 (1955).

Large plants with larger organs: *Leaves* 1−2 mm in diam. *Flowers* 25−35 mm long, with a bright yellow cup. *Perianth segments* 6−8 mm wide. *Stamens* reaching just above the middle of the perianth; anthers 5−7 mm long. *Style* 15−17 mm; stigmas mostly overtopping the anthers.

On the Orange Free State-Natal Drakensberg, to Kenya.

Vouchers: *Thode in* STE 3923, 3918 (STE); *Devenish* 1087 (PRE; K); *Devenish* 336.

48. Romulea atrandra G. J. Lewis in Flow. Pl. Afr. 14: t. 544 (1934); De Vos in Jl S. Afr. Bot. Suppl. 9: 212 (1972). Type: Cape, Laingsburg, Tweedside, *Lewis* in NBG 2703/32 (BOL, holo.!; PRE!).

Plants 100−400 mm long. *Corm* subglobose or obovoid, with sharply bent, acuminate, grooved, basal teeth bent over a rounded base. *Leaves* basal, terete to compressed cylindrical, 60−250 (−400) × 1−4 mm, bent, rigid, grooves narrow or wide. *Bracts* green with closely spaced veins, outer with narrow, inner with wide, brown-streaked, or rarely almost colourless membranous margins and tip. *Flowers* (15−) 20−40 (−45) mm long, magenta-rose to pale lilac-pink or white, usually with a violet-black blotch sometimes reduced to a few lines on each segment in the throat, cup yellow, sometimes with dark lines, outer segments greenish yellow on the backs, with dark lines and feathered veining. *Perianth tube* 4−8 mm long; segments (10−) 18−35 × (3−) 5−15 mm. *Filaments* 4−8 mm; anthers 5−10 mm long, violet or yellow. *Style* 6−10 mm; stigmas reaching halfway up the anthers or almost to their tips. *Capsules* on bent peduncles which twist spirally when dry. *Chromosome no.* 2n=22.

Found in western and south-western inland Cape districts from Vanrhynsdorp and Calvinia to Ladismith, and in southern districts from Caledon to Port Elizabeth (not in south-western Cape below the first escarpment). Map 15.

Distinguished from *R. rosea* (no. 59), which has somewhat similar flowers, in its firmer green bracts with closely spaced veins and prominent membranous margins and tips, in the sharply bent, grooved teeth of the corm tunics, and in chromosome number. Three varieties occur which differ mainly in leaf width, and size and colouring of the perianth and anthers.

1 Flowers 25−45 mm long; perianth segments 8 mm wide or wider:

  2 Widest leaf blades more than 1 mm wide with wide or narrow grooves; flowers

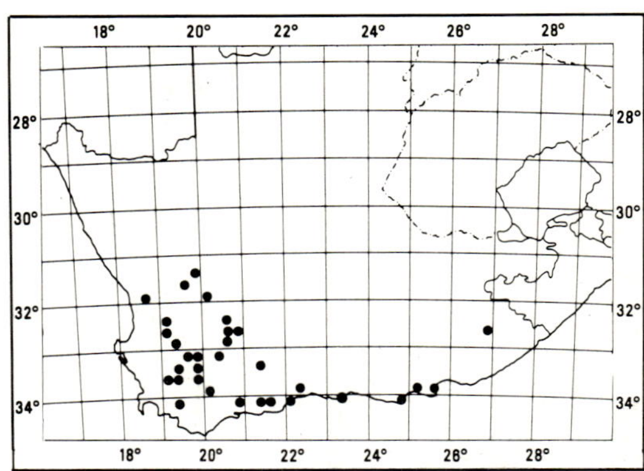

MAP 15.— **Romulea atrandra**

    mostly magenta-rose with large dark blotches in the throat .... (a) var. *atrandra*

  2 Widest leaf blades 1 mm or less in diam. with narrow grooves; flowers paler: pink, lilac or white with small or sometimes large dark blotches in the throat or without blotches ....(b) var. *esterhuyseniae*

1 Flowers up to 25 mm long or rarely slightly longer; perianth segments 3−7 mm wide .... (c) var. *lewisiae*

(a) var. **atrandra.**

Widest leaves more than 1 mm in diam. with wide or narrow grooves. *Bracts* with brown-streaked or almost colourless membranous margins and tips. *Flowers* 25−45 mm long, mostly magenta-rose with a dark blotch on each segment, cup yellow or orange-yellow, usually with dark violet longitudinal lines. *Anthers* dark purple or yellow, reaching about halfway or less up the perianth.

On inland plateaux from Ceres to Sutherland, Laingsburg, Worcester and Caledon.

Vouchers: *Loubser* 2073 (NBG); *De Vos* 1595 (STE); *Acocks* 16976 (PRE; K); *Hall* 3246 (NBG).

(b) var. **esterhuyseniae** *De Vos* in Jl S. Afr. Bot. Suppl. 9: 215, fig. 69 (1972). Type: Cape, Ladismith, Towerkop on the Swartberge, *Esterhuysen* 13924 (BOL, holo.!; PRE!; SAM!; STE!; K!).

This differs from var. *atrandra* in the following: *Leaves* not more than 1 mm in diam., with narrow grooves. *Bracts* with brown-streaked membranous margins. *Flowers* paler magenta to pale lilac or white,

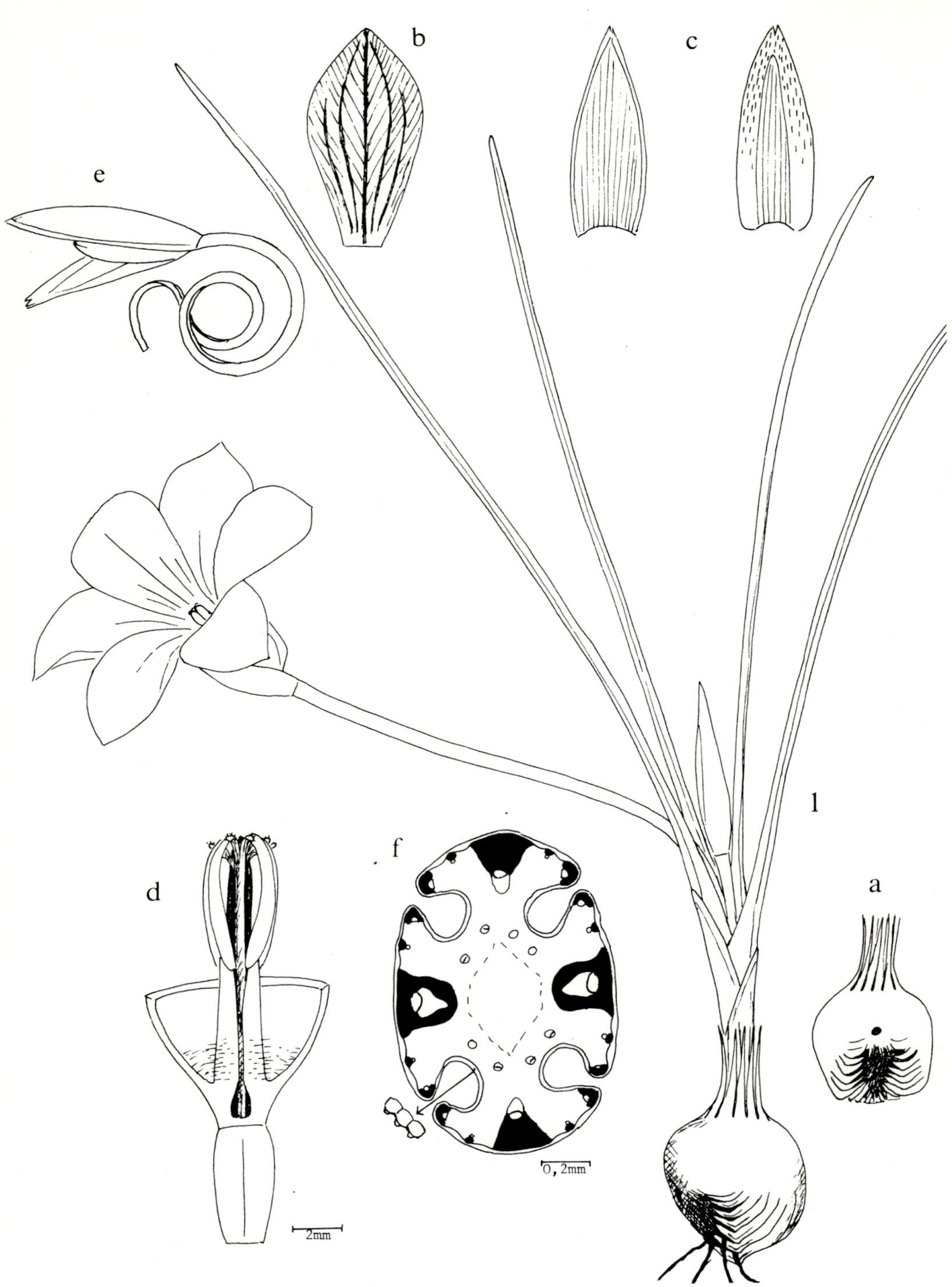

b
c
e
d
f
a
1
2mm
0,2mm

with a dark blotch or sometimes dark lines, or with a pale transverse zone on each segment. *Anthers* mostly yellow, sometimes with dark lines. Fig. 14.

Mainly on stony mountain slopes from Vanrhynsdorp to Ladismith and in the districts of Oudtshoorn, Riversdale and Port Elizabeth.

Vouchers: *Esterhuysen* 14568, 18719 (BOL; STE); *Barnes* in BOL 19470 (BOL); *Wurts* 1161 (NBG); *Taylor* 5016 (PRE; K).

(c) var. **lewisiae** *De Vos* in Jl S. Afr. Bot. Suppl. 9: 217 (1972). Type: Cape, Port Elizabeth, Victoria Park Lands, *Long* 617 (PRE, holo.!; GRA!; K!).

Leaves up to 1 mm in diam. *Bracts* 9—15 mm long, with brown-streaked membranous margins. *Flowers* 15—25 mm long, pale lilac-pink to white, outer segments on the backs with 3 dark veins. *Perianth segments* 3—7 mm wide. *Anthers* reach more than halfway up the perianth.

From Montagu and Riversdale to Knysna, near Port Elizabeth, and on mountains near Alice and Clanwilliam.

Vouchers: *Muir* 949 (BOL); *Hanekom* 1229 (STE); *Batten* 2—Pl 62 (NBG); *Denman* 43 (GRA).

48a. **Romulea vlokii** *De Vos*, sp. nov. R. atrandram *proxima sed cormo tunicis dentibus rectis acuminatis ad fundum acutum cormi convergentibus differt.*

Type: Cape, Oudtshoorn, Kammanassie, southern slopes of Buffelsberg near Diepkloof, *Vlok* 384 (STE, holo.!).

Plants 270—350 mm long. *Corm* obovoid, tunics with straight acuminate teeth converging to a pointed base. *Leaves* filiform, 150—340 xc. 1 mm, grooves narrow. *Peduncles* 70—150 mm long. *Bracts* firm, green, with prominent closely spaced veins, outer with narrow, inner with wide, papery, brown-streaked margins and tips. *Flowers* 32—40 mm long, pink, cup orange-yellow, outer perianth segments with purple stripes on backs. *Perianth tube* short; segments 22—25 x 8—12 mm. *Filaments* c. 4—5 mm; anthers c. 7 mm long, yellow, reaching less than halfway up the perianth

segments. *Style* c. 10 mm, pale; stigmas reaching halfway up the anthers.

Found in the Kammanassie region of the Oudtshoorn district, in moist sandy soil (3322—DB).

Voucher: Type only.

This rare species, found only recently (27-7-1982), has bracts similar to the brown-streaked bracts of *R. atrandra* (no. 48) and *R. luteoflora* (no. 49). The flower and leaf anatomy approximate that of *R. atrandra* var. *esterhuyseniae*, (no. 48b) but the corm is similar to that of *R. cruciata* no. 60) and *R. eximia* (no. 61).

49. **Romulea luteoflora** *(De Vos) De Vos* in Jl S. Afr. Bot. Suppl. 9: 210 (1972). Type: Cape, Ceres, top of Theronsberg Pass, *De Vos* 1570 (STE, holo.!).

Very closely related to *R. atrandra* (no. 48) from which it differs as follows: *Leaves* 0,5—1(—2) mm in diam. *Inner bract* sometimes slightly longer and wider than the outer. *Flowers* buttercup-yellow, usually with a dark reddish brown blotch or three dark lines on each segment, outer segments with brownish black lines and feathered veining or speckling on the backs. *Perianth tube* often cup-shaped. *Anthers* (3—)6—9 mm long, dark brown or sometimes yellow. *Chromosome no.* 2n=20.

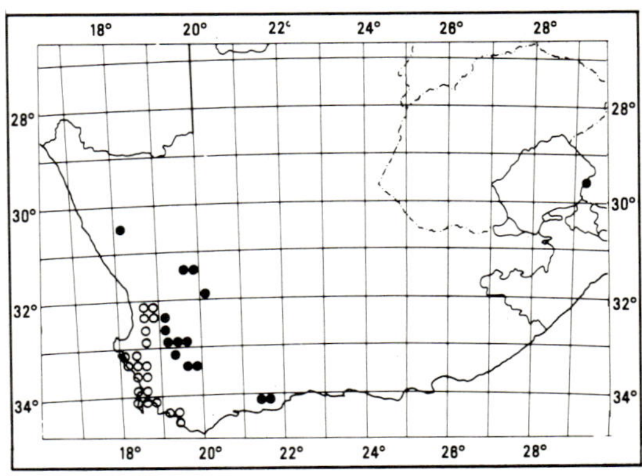

MAP 16.—● **Romulea luteoflora**
     ○ **R. obscura**

FIG. 14.—1, **Romulea atrandra** var. **esterhuyseniae**, habit, × 1; a, corm seen from the opposite side; b, outer perianth segment, outer face; c, outer (left) and inner (right) bract; d, perianth tube, stamens and pistil; e, mature capsule; f, transverse section of leaf (*De Vos* 2101).

With a wide, apparently disjunctive, inland range: on the Kamiesberg, Namaqualand, and from Calvinia to Clanwilliam, Ceres and Worcester, also near Riversdale and on the Sani Pass, Lesotho. Map 16.

Distinguished by its corm, firm green bracts with prominent membranous margins and tips, and flower colouring; at first thought to be a variety of *R. atrandra* (no. 48). Two varieties are now recognized after the recent discovery of a small montane form on the Lesotho-Natal border. (See descriptions below for distinguishing characters).

(a) var. **luteoflora.**

R. atrandra G. J. Lewis var. *luteoflora* De Vos in Flower. Pl. Afr. 29: t. 1135 (1952).

Peduncle 40—140 mm long. *Flowers* 25—40 (—45) mm, often with a dark reddish brown blotch on each segment. *Anthers* 6—9 mm long.

From the Kamiesberg, Calvinia to Worcester, and Riversdale.

Vouchers: *Leighton* 1288 (BOL); *Salter* 4708 (BOL; K); *De Vos* 1615 (STE); *Marloth* 6137 (PRE; STE); *Lewis* 2666 (SAM).

(b) var. **sanisensis** *De Vos*, var. nov.

A varietate *luteoflora* pedunculis brevioribus, bracteis et floribus parvioribus bene distinguitur.

Type: Lesotho, Sani Pass, *Hilliard & Burtt* 7153 (STE, holo.!; NU).

Differs from the typical variety in its smaller size: *Corm* small, with softer, more membranous tunics. *Peduncle* 25—30 mm long. *Flowers* 17—22 mm long, just above ground level. *Perianth segments* 7—12 × 4—5 mm, without dark blotches in the throat. *Anthers* 3 mm long.

From Sani Pass, Lesotho, W of the border post, on flat grassland.

Voucher: *Hilliard & Burtt* 7153 (STE; NU).

50. **Romulea hallii** *De Vos* in Jl S. Afr. Bot. Suppl. 9: 217, fig. 70 (1972). Type: Cape, Sutherland, summit of Verlatekloof, *Hall* 3176 (NBG, holo.!).

Closely related to *R. atrandra* (no. 48) from which it differs in the following: *Leaves* compressed cylindrical, 100—130 × 2—3 mm. *Outer bract* with a green, almost triangular, central basal part and wide, minutely speckled membranous margins and tip; inner green or greenish in the centre, margins and tip as the outer. *Flowers* 22—33 mm long, pale wistaria-blue, each segment with a violet and below that an almost black blotch, cup and throat orange-yellow with dark lines, outer segments on the backs with; 3—5 violet veins and fine feathered veining. *Perianth tube* 5—6 mm long; segments 15—22 × 8—10 mm. *Filaments* 5—6 mm; anthers 4—5 mm long, yellow. *Style* 10—12 mm; stigmas at or just below the anther tips. *Capsules* on strongly recurved, and later flexuose peduncles. *Chromosome no.* 2n = 22.

Found only on the Great Roggeveld plateau south-west of Sutherland near the summit of Verlatekloof, on clayey ground, at c. 1 500 m altitude (3220—BC, —DA).

Vouchers: *De Vos* 2215 (STE; PRE); *Stayner* s.n., 8—7—1968 (NBG); *Hall* 3176 (NBG).

An early flowering species (May to July), with characters intermediate between *R. atrandra* (no. 48) and *R. komsbergensis* (no. 51). It was at first thought to be a hybrid between these two species; it produces, however, well developed seeds.

51. **Romulea komsbergensis** *De Vos* in Ann. Univ. Stell. 28A, 3: 69, fig. 4 (1952); in Jl S. Afr. Bot. Suppl. 9: 219, figs 71 & 76 (1972). Type: Cape, Sutherland, plateau N of Komsberg Pass, *De Vos* 1582 (STE, holo.!).

Plants 120—300 mm long. *Corm* tunics with grooved, bent, basal teeth. *Leaves* basal, filiform, usually arcuate, 120—300 × c. 1 mm, with the adaxial groove often open up to 30 mm from the tip. *Bracts* green in the centre of the upper half, often submembranous below, with wide, mostly fawn-coloured membranous margins and tips. *Flowers* 20—35 mm long, rosy-magenta, with a narrow blue transverse band, cup buttercup-yellow with a brown base, outer segments with 5—7 violet veins and fine feathered veining, or reddish purple or irregularly marked. *Perianth tube* 3—4 mm long; segments usually obtuse, 15—28 × 8—15 mm. *Filaments* 4 mm; anthers 3—5

FIG. 15.—1, **Romulea komsbergensis,** habit, × 7/8; 1a, perianth tube, stamens and pistil; 1b, outer perianth segment, lower face; 1c, bracts, outer (left) and inner (right); 1d, mature capsule (*De Vos* 1582). 2, **R. multifida,** habit, × 7/8; 2a, perianth tube, stamens and pistil; 2b, perianth segment, upper face; 2c, outer (left) and inner (right) bract; 2d, almost mature capsule (STE 27158).

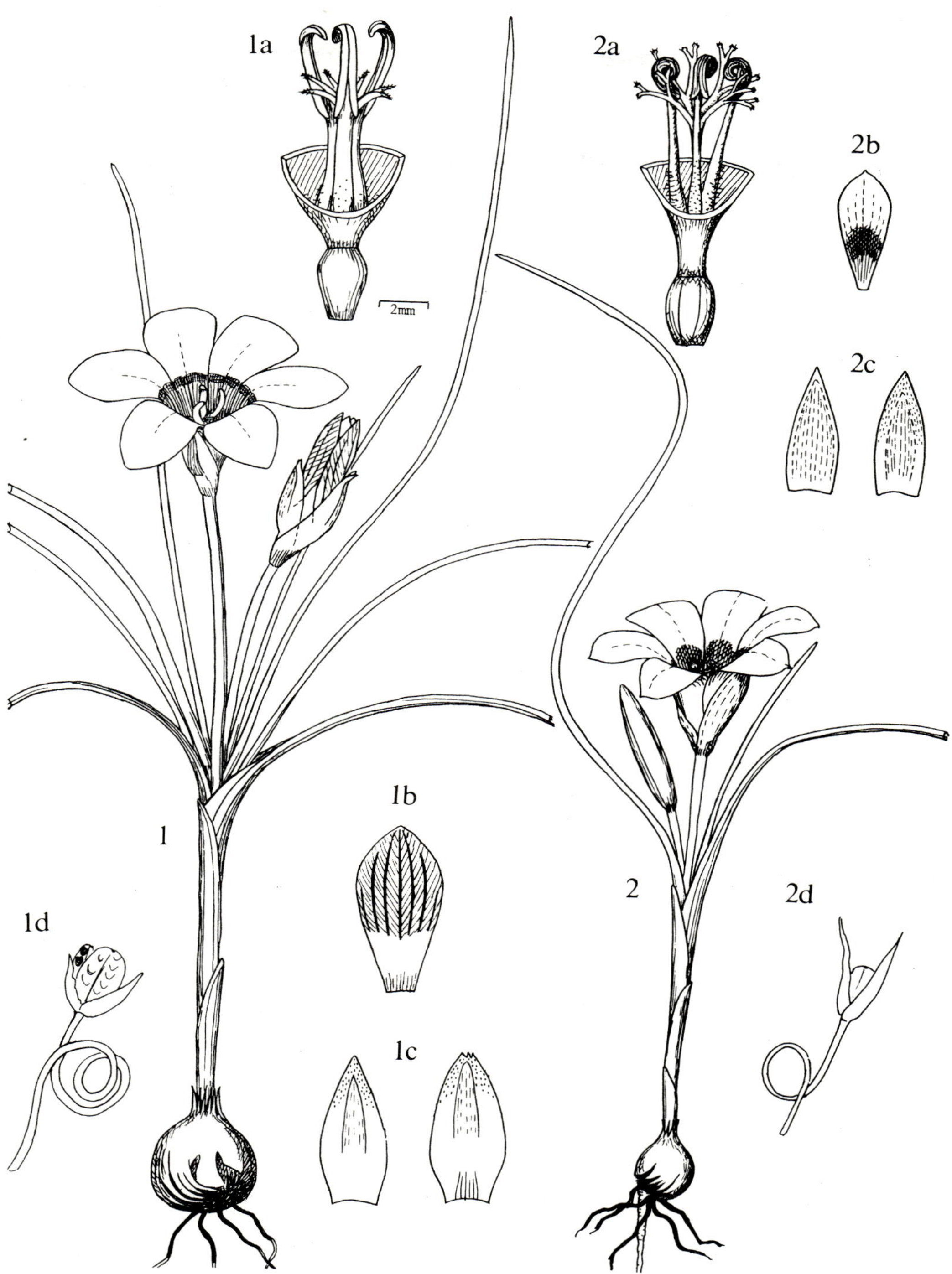

1a
2a
2b
2c
2mm
1b
1
1d
1c
2
2d

mm long, violet, later yellow with incurved tips, pollen brown or rust-coloured. *Style* 5−7 mm with acuminate, often violet branches; stigmas minute, terminal, with a tuft of long papillae. *Capsule* on a spirally coiled peduncle when dry. *Chromosome no.* 2n=20. Fig. 15:1.

Found only on the Great Roggeveld plateau SSE of Sutherland, towards Komsberg Pass, at c. 1500 m altitude, in sandy soil (3220−DA, DB).

Vouchers: *De Vos* 1932 (STE); *Leistner* 277; *Acocks* 18451 (BOL; M).

Distinguished from *R. atrandra* (no. 48) by its less rigid, more membranous bracts, incurved or circinnate anthers, acuminate style branches, and perianth cup with a brown base. One or two of the style branches are sometimes split, thus forming more than six stigmas.

52. **Romulea multifida** *De Vos* in Ann. Univ. Stell. 28A, 3: 71, fig. 5 (1952); in Jl S. Afr. Bot. Suppl. 9: 221, figs 72 & 78 (1972). Type: Cape, plateau W of Sutherland, *Joubert* in STE 27158 (STE, holo.!).

Very closely related to *R. komsbergensis* (no. 51) from which it differs in the following: Plants up to 200 mm long. *Leaves* 100−200 × 0,5−1 mm, with the adaxial groove closed nearer the base. *Bracts* submembranous, the inner with wide membranous margins which are colourless towards the base and fawn upwards, and a membranous tip. *Flowers* with a brown or darker blotch on each segment in the throat, cup yellow, outer segments on the backs with 3−5 violet veins. *Perianth tube* 5−7 mm long; segments minutely apiculate, 14−20 × 6−9 mm. *Filaments* 5−6 mm; anthers circinnate, purplish or yellowish, pollen yellow. *Style* branches multifid, violet; stigmas 12 or more, small, terminal. *Chromosome no.* 2n=22. Fig. 15:2.

Found only on the Great Roggeveld plateau west and south of Sutherland, at c. 1 500 m altitude, in sandy soil (3220−BC, DA).

Vouchers: *Burchell* 1321 (K); *De Vos* 2194 (STE); *Joubert* in STE 27158.

Distinguished by its multifid style branches and numerous minute terminal stigmas and by its inrolled anthers.

53. **Romulea malaniae** *De Vos* in Jl S. Afr. Bot. Suppl. 9: 223, fig. 73 (1972). Type: Cape, Montagu, Sandvlei 23 km S of Matroosberg Station, *Malan* in STE 30312 (STE, holo.!).

Plants 120−250 mm long. *Corm* subglobose, with bent, acuminate, grooved basal teeth. *Leaves* few, basal, filiform, 120−250 × c. 1 mm, grooves narrow. *Bracts* largely membranous or greenish in the centre of the upper half, often with purplish veins, with membranous margins and tips. *Flowers* 15−32 mm long, pale yellow, outer segments brownish on the backs. *Perianth tube* 5−9 mm long, narrow in the lower half, funnel-shaped towards the top; segments 8−20 × 4−5 mm. *Filaments* 4−5 mm; anthers 3−4 mm long. *Style* 8−10 mm; stigmas generally below the anther tips. *Capsules* often pushed underground by the strongly recurved peduncles which later coil up when dry. *Chromosome no.* 2n=24.

Found only once in the Montagu district between Matroosberg Station and Koo, in sandy pockets at the foot of a stony ridge (3320−DB).

Vouchers: *Malan* in STE 30312 (STE).

Related to the *R. atrandra* (no. 48) group of species and distinguished by its small, pale yellow flowers on rather stout peduncles which recurve with an almost knee-like bend even before the flowers fade, often pushing the developing capsules underground. The dry capsule again appears above ground when the peduncle coils up.

54. **Romulea diversiformis** *De Vos* in Ann. Univ. Stell. 28A, 3: 63, fig. 2 (1952); in Jl S. Afr. Bot. Suppl. 9: 225, fig. 74 (1972). Type: Cape, Sutherland, Damslaagte, Klein Roggeveld, *De Vos* 1581 (STE, holo.!).

Plants 80−200 mm long. *Corm* subglobose, with bent, acuminate, grooved basal teeth. *Leaves* several, basal, filiform, 70−200 × 0,5−1,5 mm, grooves narrow. *Bracts* green or greenish, outer with narrow, inner with wide membranous margins and membranous tips. *Flowers* 25−35 mm long, buttercup-yellow. *Perianth tube* 4−6 mm long; segments 18−28 mm long, the outer obovate to oblanceolate, 7−12 mm wide, brownish on the backs, the inner 10−16 mm wide, obovate-cuneate, sometimes with flexuose margins. *Filaments* 4,5−6 mm; anthers 5−7,5 mm long, slightly spreading, with incurved tips, reaching about halfway up the perianth. *Style* 13−17 mm long; stigmas narrowly ligulate, reaching some millimetres above the anther tips. *Capsules* on bent peduncles. *Chromosome no.* 2n=20.

From the Sutherland district, in moist localities (3220—BC, DA, DC).

Vouchers: *Joubert* in STE 27157; *De Vos* 1929 (STE); *Acocks* 18452 (PRE; K; M), & 17173 (PRE; K); *Leistner* 279.

Related to the *R. atrandra* group of species and distinguished by its bright yellow flowers with the inner perianth segments several millimetres wider than the outer, and by long styles with stigmas above the anther tips.

55. **Romulea membranacea** *De Vos* in Jl S. Afr. Bot. Suppl. 9: 227, fig. 75 (1972). Type: Cape, Calvinia near Elandsfontein, 29 km, NW of Middelpos, *Salter* 3488 (BOL, holo.!; BM!; K!).

Plants 70—120 mm long. *Corm* with almost straight, acuminate basal teeth converging to a pointed base. *Leaves* basal, filiform, recurved or flexuose, 80—100 × c. 0,5 mm, glabrous or minutely ciliate, grooves narrow. *Bracts* largely membranous, sometimes submembranous in the middle, with wide, brown-speckled membranous margins and tip. *Flowers* 20—30 mm long, shiny golden-yellow, with 1—3 slender dark veins on each segment, outer segments blotched or striped with brown or purple on the backs. *Perianth tube* 3—5 mm long; segments 15—22 × 4—7 mm. *Filaments* 5—6 mm; anthers 3—4 mm long. *Style* 9—10 mm; stigmas at or near the anther tips. *Capsule* subglobose, on strongly recurved, twisted peduncles. *Chromosome no.* 2n=24.

The only known site is in the Calvinia district north-west of Middelpos, on red sandy ground (3119—DD).

Vouchers: *De Vos* 2221 (STE; PRE); *Salter* 3488 (BOL; BM; K).

Distinguished by its corm with a pointed base, very slender flexuose or recurved leaves, bright yellow flowers and bracts which are largely membranous. The species is seemingly not closely related to any other species of *Romulea*. It may perhaps form a link between the *R. atrandra* (no. 48) and *R. cruciata* (no. 60) species groups.

56. **Romulea obscura** *Klatt* in Abh. Naturforsch. Ges. Halle 15: 399 (1882); in Dur. & Schinz, Consp. Fl. Afr. 5: 165 (1895); Bak. in F.C. 6: 40, pro syn.; De Vos in Jl S. Afr. Bot. Suppl. 9: 231, figs 64, 65, 82, 83, 84. Type: Cape, without locality, *Drège* 4041 (B, holo.†; S, lecto.!; BM!).

Plants 100—500 mm. *Corm* subglobose, with strong, curved basal teeth bent over a rounded base. *Stem* short or up to 80 mm long, usually hidden under leaf bases. *Leaves* basal or sometimes basal and cauline, 100—500 × 0,5—1 mm. *Bracts* green or brownish green, inner with wide brown or speckled or sometimes colourless membranous margins. *Flowers* 15—50 mm long, deep old-rose to apricot or yellow, sometimes with dark blotches, cup greenish yellow or orange-yellow, outer segments on the backs brownish purple or irregularly marked or striped. *Perianth tube* 2—5 mm long; segments 10—40 × 3—12 mm. *Filaments* 3—8 mm; anthers 3—9 mm long. *Style* 6—15 mm. *Capsules* on strongly curved, flexuose, widely patent peduncles. *Chromosome no.* 2n=22.

Generally in the south-western Cape from Hopefield to the Cape Peninsula, with outliers to Clanwilliam and to Caledon and Bredasdorp, on sandy or loamy flats at low altitude. Map 16.

A large, polymorphic species in which four largely interfertile varieties are recognized, with some intermediates between the varieties. The flower colouring, which varies from deep old-rose to yellow, without any bluish tints, distinguishes this species from *R. rosea* (no. 59); also the recurved flexuose peduncles and the 2n chromosome number of 22.

1 Perianth yellow or apricot or rarely pale pink, rarely with dark blotches and flowers then yellow:

  2 Flowers 15—25 (—35) mm long; anthers 3—5 mm long .... (a) var. *obscura*

  2 Flowers 25—45 mm long, rarely with dark blotches in the throat; anthers 7—9 mm long ....(b) var. *campestris*

1 Perianth deep rosy-pink or dark old-rose almost terracotta, generally with dark blotches in the throat:

  3 Flowers usually 18—25 mm long; perianth segments 4—7 mm wide, with small purplish blue or greyish blue blotches in the throat........................ (c) var. *subtestacea*

  3 Flowers (25—) 30—50 mm long; perianth segments 7—12 mm wide, with dark red or purple blotches in the throat, usually on a violet-blue or grey background .... (d) var. *blanda*

(a) var. **obscura.**

R. parviflora Eck., Top. Verz. 19 (1827), nom. nud., non (Salisb.) Britten (1914), (see Nordenstam in Jl S. Afr. Bot. 38: 282 (1972) and De Vos in Jl S. Afr. Bot. 39: 111 (1973)). *Trichonema parviflorum* Steud., Nom. Bot. edn 2, 2: 702 (1841), nom. nud.

R. elegans Klatt var. *parviflora* sensu G. J. Lewis in Adamson & Salter, Fl. Cape Penins. 223 (1950), non Bak.

Stem short. *Leaves* basal. *Flowers* 15−25 (−35) mm long, yellow or apricot, sometimes with 1−3 dark veins on each segment. *Perianth* segments 3−5 mm wide. *Style* 6−10 mm long.

From Hopefield to Bredasdorp.

Vouchers: *Esterhuysen* 19004 (BOL); *Gillett* 1038 (BOL); *Malan* s.n. in STE 30203; *De Vos* 1470 (STE); *Bolus* 4601 (BM; K).

Distinguished from *R. rosea* var. *australis* (no. 59c) by its more erect, narrower leaves and flower colouring. Some indifferent herbarium specimens can, however, be confused.

(b) var. **campestris** *De Vos* in Jl S. Afr. Bot. Suppl. 9: 237, figs 65 & 83 (1972). Type: Cape, between Clanwilliam and Graafwater, *De Vos* 1897 (STE, holo.!).

Stem short or sometimes up to 80 mm long. *Leaves* basal and sometimes also cauline, with wide leaf sheaths. *Flowers* 25−45 mm long, varying from yellow to apricot or pale rosy-pink, rarely with dark blotches (in yellow flowers), cup golden-yellow with slender dark lines. *Perianth segments* 5−10 mm wide. *Style* 8−15 mm long.

From Clanwilliam to Malmesbury.

Vouchers: *Salter* 3680 (BOL); *Lewis* 5238 (NBG; STE); *Barker* 4764 (NBG; STE); Marsh 783 (STE; PRE).

(c) var. **subtestacea** *De Vos* in Jl S. Afr. Bot. Suppl. 9: 238 (1972). Type: Cape, Malmesbury, Soutpan NE of Ysterfontein, *De Vos* 2037 (STE, holo.!).

Stem short or up to 30 mm long. *Leaves* mostly basal. *Flowers* 18−25(−28) mm long, dark old-rose or almost terracotta, with a small purplish blue or greyish blue blotch on each segment, cup greenish yellow with dark lines. *Perianth segments* −7 mm wide. *Style* 7−9 mm long.

From Hopefield to the Cape Peninsula.

Vouchers: *De Vos* 2037 (STE); *Salter* 3003 (BOL); *Bolus* 22897 (BOL); *Lewis* 1057 (SAM).

This variety stands intermediate between var. *obscura* and var. *blanda*, with its flowers coloured as in var. *blanda*, except that the blotches are smaller and not encircled by a differently coloured zone. In the Kraaifontein area near Cape Town intermediates occur between var. *subtestacea* and var. *obscura*.

(d) var. **blanda** *De Vos* in Jl S. Afr. Bot. Suppl. 9: 239, figs 64 & 84 (1972). Type: Cape, Malmesbury, between Mamre and Ysterfontein, *Bolus* 20725 (BOL, holo.!; K!).

Stem short or up to 30 mm long. *Leaves* mostly basal. *Flowers* (25−)30−50 mm long, deep rosy-pink or dark old-rose, with a maroon or purple-black blotch frequently encircled by a bluish or greyish zone on each segment, cup orange or golden-yellow with slender dark lines. *Perianth segments* 6−12 mm wide. *Style* 10−15 mm long.

From Vanrhynsdorp to the Cape.

Vouchers: *Barker* 10394 (NBG); *De Vos* 2085 (STE); *Lewis* 1062 (SAM) & 5535 (NBG); *Bolus* 24785 (BOL).

This variety is distinguished from the sympatric *R. eximia* (no. 61) and *R. hirsuta* (no. 21) by its round-based corm and the distinctive colouring of the blotches on the perianth; from *R. eximia* also by its orange-yellow cup and the curvature of the dried peduncles; and from *R. hirsuta* by its less green bracts.

57. **Romulea monticola** *De Vos* in Jl S. Afr. Bot. Suppl. 9: 241, fig. 85 (1972). Type: Cape, Calvinia, summit of Vanrhyn's Pass, *De Vos* 1924 (STE, holo.!).

Plants 100−250 mm long. *Corm* sub-globose, with strongly curved basal teeth bent, and often broken over a rounded base. *Leaves* basal, filiform, sometimes minutely ciliate on rib margins, 90−250 × 0,5−1 mm, grooves narrow. *Bracts* greenish or brownish purple, inner with wide, brown, membranous margins and tip. *Peduncles* reddish brown, subterete. *Flowers* 22−35 mm long, golden-yellow, frequently with a darker yellow cup and one or more dark veins on each segment, outer segments reddish brown on the backs or with 3−5 dark veins. *Perianth tube* 3−4 mm; segments 18−27 × 5−7 mm. *Filaments* 4−5 mm; anthers 3−5 mm, at first lightly joined at the tips, reaching less than halfway up the

---

FIG. 16.−1, **Romulea monticola**, habit, × 1; 1a, outer perianth segments of two plants, lower faces; 1b, outer (left) and inner (right) bract; 1c, transverse section of leaf (*De Vos* 1924). 2, **R. rosea** var. **rosea**, habit, × 1; 2a, outer perianth segment, lower face; 2b, bracts, outer (left) and inner (right); 2c perianth tube, stamens and pistil (*De Vos* 1775).

1a
1b
1
1c
2
2a
2b
2c
0,2mm
2mm

perianth. *Style* 7−10 mm; stigmas at or just below the anther tips. *Capsules* on suberect or curved peduncles. *Chromosome no.* 2n=22. Fig. 16:1.

Found above the Vanrhyn's Pass on the Bokkeveld Mountains plateau between Calvinia and Vanrhynsdorp, also on the Gifberg plateau, Vanrhynsdorp, on sandy or loamy ground at 700−800 m altitude (3118−DC; 3119−AC)

Vouchers: *Acocks* 18212; *Bond* 1194 (NBG); *Lewis* 5844 (NBG); *Compton* 20883 (NBG); *Esterhuysen* 5274 (BOL).

Closely related to *R. obscura* (no. 56) and distinguished chiefly by its bright golden-yellow flowers on brownish subterete peduncles which remain suberect or bend only slightly in the fruiting stage. The flower can be confused with that of the sympatric *R. montana* (no. 11) but *R. monticola* differs in its corm, less green bracts, leaf anatomy with fibre bundles along the margins of the leaf ribs, and in chromosome number.

58. **Romulea cedarbergensis** *De Vos* in Jl S. Afr. Bot. Suppl. 9: 245, fig. 86 (1972). Type: Cape, Clanwilliam, Wolfsberg in the Cedarberg Mtns, *De Vos* 2030 (STE, holo.!).

Plants 35−180 mm long. *Corm* subglobose, 3−6 mm in diam., with curved basal teeth bent over a rounded base. *Leaves* few, basal, filiform, 30−180 × 0,3−0,5 mm, with very narrow grooves. *Bracts* submembranous, purplish or greenish, inner with wide, fawn or brown-speckled or edged membranous margins. *Flowers* often single, 15−25 mm long, white or pale pink with a golden-yellow cup, outer segments with purplish blue stripes on the backs. *Perianth tube* 3−5 mm long, narrow in the lower half; segments 7−16 × 2,5−6 mm. *Filaments* 4−6 mm, orange-yellow; anthers 2−3,5 mm pale yellow. *Style* 7−9 mm; stigmas at or just above the anther tips. *Capsules* on suberect peduncles.

Found only on high mountain plateaux of the Cedarberg range, in the region of Clanwilliam, at c. 1 500 m altitude, in shallow hollows (3219−AA, CA, DC).

Vouchers: *Esterhuysen* 8043 (BOL); *Stokoe* in SAM 63691 (SAM); *De Vos* 2030 (STE).

A high mountain species with few, small flowers and very narrow leaves, nearest the *R. obscura* (no. 56) and *R. monticola* (no. 57) group of species.

59. **Romulea rosea** *(L.) Eckl.*, Top. Verz. 19 (1827); Bak. in J. Linn. Soc., Bot. 16: 88 (1877); Handb. Irid. 103 (1892), excl. all vars; in F. C. 6: 41 (1896), excl. all vars; Klatt in Abh. Naturforsch. Ges. Halle 15: 400 (1882); Bég. in Malpighia 23: 60 (1909), excl. syn. *Ixia fugax, R. vulgaris* and vars 3−7; G. J. Lewis in Adamson & Salter, Fl. Cape Penins. 222 (1950); De Vos in Jl S. Afr. Bot. Suppl. 9: 246 (1972). Iconotype: Mill., Fig. Pl. 160: t. 240 (1760).

Plants 150−600 mm long. *Corm* subglobose, with curved acuminate basal teeth bent over a rounded base. *Leaves* basal, filiform to compressed cylindrical, 0,5−2,5 mm in diam. *Bracts* greenish or purplish, the inner with wide, brown or brown-streaked membranous margins. *Flowers* 15−48 mm long, magenta, pink, lilac-pink or white, often with a violet-blue throat, cup pale yellow to orange-yellow, outer segments variously coloured or marked on the backs. *Perianth tube* 2−8 mm long; segments 10−38 × 3−10 mm. *Stamens* 7−16 mm long; anthers 3−10 mm long, pale to golden-yellow, subequal to or longer than the filaments. *Style* 7−18 mm; stigmas below, above or at the anther tips. *Capsules* on peduncles which curve after flowering and straighten later. *Chromosome no.* 2n=18.

Widely distributed from Vanrhynsdorp to the Cape and to Port Elizabeth; var. *australis* also naturalized in Australia and found on St Helena, Tristan da Cunha and the Channel Islands. Map 17.

A large, polymorphic species comprising five varieties and several forms, often connected by intermediates. Distinguished by its rounded corm, short stem, greenish or purplish bracts, small to large flowers usually in several shades of pink, with a yellow

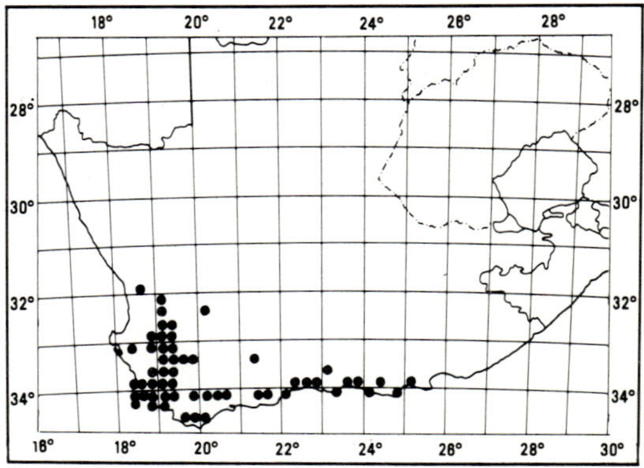

Map 17.— **Romulea rosea**

to orange-yellow cup and often a violet-blue throat. Several of the specimens cited in F. C. have been relegated to other species.

1 Flowers white with a golden-yellow or orange-yellow cup; outer segments reddish on the backs, irregularly blotched or with a pale median line .... (e) var. *elegans*

1 Flowers magenta to pink or white; if white then cup pale yellow and outer segments yellowish or greenish on the backs:

  2 Stigmas usually overtopping the anthers; corm slightly depressed globose with somewhat membranous, not very hard tunics .... (a) var. *rosea*

  2 Stigmas at or below the anther tips; corm subglobose with outer tunics hard and rigid:

    3 Widest leaf blades 0,5−1 mm in diam., filiform; outer perianth segments with violet veins and fine feathered veining or blotched or purple, sometimes with a pale median zone .... (b) var. *reflexa*

    3 Widest leaf blades 1−2 mm in diam., compressed cylindrical; outer perianth segments yellowish green or sometimes with 3−5 dark lines and flowers then small:

      4 Flowers 15−22(−25) mm long; perianth segments up to 4 mm wide; outer segments with stripes or yellowish green ....(c) var. *australis*

      4 Flowers (22−)25−35 mm long; perianth segments 4−8 mm wide; outer segments usually yellowish green on the backs ....(d) var. *communis*

(a) var. **rosea.**

Ixia rosea L., Syst. Nat. 75 (1767); Ait., Hort. Kew. 1: 56 (1789); Willd., Sp. Pl. 1: 196 (1797); Pers., Syn. Pl. 1: 46 (1805); Vahl, Enum. Pl. 2: 49 (1806), excl. syn. Lam.; Roem. & Schult., Syst. Veg. 1: 375 (1817), partly, excl. syn. Lam., Red. and F. Delaroche.

Ixia bulbocodium Murray, Syst. Veg. edn 13, 75 (1774), partly, excl. cit. Clus.; Thunb., Diss. Ixia 6 (1783); Fl. Cap. edn 2, 55 (1823) var. γ; Lam., Encycl. 3: 335 (1789) var. δ. *Trichonema roseum* (L.) Ker-Gawl. in Curtis's bot. Mag. 30: t. 1225 (1809), partly, excl. syn. Lam., Red., Burm. and F. Delaroche; Steud., Nom. Bot. edn 2, 2: 702 (1841), partly, excl. syn. Burm. and Lam.; Klatt in Linnaea 34: 663 (1865−66), partly, excl. syn.; non Spreng. (1825). *Bulbocodium roseum* (L.) Kuntze, Rev. Gen. 2: 701 (1891).

Crocus capensis Burm. f., Prodr. Fl. Cap. 2 (1768), excl. var. *floribus luteis*. Type: *Burman* Herb. (G, holo.!).

Ixia chloroleuca Jacq., Icon. Pl. Rar. 2: t. 272 (1789); Coll. 4: 180 (1790); Willd., Sp. Pl. 1: 196 (1797); Pers., Syn. Pl. 1: 46 (1805); Roem. & Schult., Syst. Veg. 1: 373 (1817). *Trichonema ochroleucum* Ker-Gawl. in König & Sims, Ann. Bot. 1: 223 (1805); Klatt in Linnaea 34: 670 (1865−66). *Ixia ochroleuca* Vahl, Enum. Pl. 2: 50 (1806); Poir., Encycl. Suppl. 3: 200 (1813), partly, excl. var. β, var. γ and syn. Lam. *Trichonema chloroleucum* (Jacq). Ker-Gawl. in Curtis's bot. Mag. 30: sub t. 1225 (1809); Irid. Gen. 82 (1827). *R. chloroleuca* (Jacq.) Bak. in J. Linn. Soc., Bot. 16: 89 (1877); in F. C. 6: 42 (1896); Klatt in Abh. Naturforsch. Ges. Halle 15: 398 (1882). *R. rosea* var. *chloroleuca* (Jacq.) Bég. in Malpighia 23: 62 (1909). Iconotype: Jacq., Icon. Pl. Rar. 2: t. 272 (1789).

R. rosea var. *celsii* Planch., Fl. Serres 8: t. 799 (1852−53). *R. celsii* (Planch.) Klatt in Abh. Naturforsch. Ges. Halle 15: 400 (1882). Iconotype: Planch., Fl. Serres 8: t. 799 (1852−53), lecto.!.

Corm with outer tunics somewhat membranous, split irregularly into rather membranous, bent, basal teeth. *Leaves* filiform, c. 1 mm in diam. *Flowers* 22−35(−45) mm long, magenta or rosy-pink to white, sometimes with a violet-blue zone in the throat, outer segments greenish mauve or blotched on the backs. *Style* 12−18 mm long; stigmas usually overtopping the anthers by several millimetres. Fig. 16: 2.

From Piketberg to Caledon.

Vouchers: *Salter* 4955 (BOL); *Barker* 174 (NBG); *Schlechter* 9124 (BOL; GRA; PRE; BM; G; K); *De Vos* 1472 (STE; PRE).

This variety is distinguished mainly by its long styles. White-flowered specimens were previously described as *R. chloroleuca* and *Trichonema ochroleucum*. Flowering period late, mainly in November.

(b) var. **reflexa** *(Eckl.) Bég.* in Annu. Conserv. Jard. bot. Genève 11−12: 158 (1908); in Malpighia 23: 64 (1909). Type: Cape, am Löwenberg u. Grünpoint, (Cape Town, Lion's Head and Green Point) *Ecklon* s.n., 10 Sept. 1826 (S, lecto.!).

R. reflexa Eckl., Top. Verz, 18 :1827), excl. syn. Thunb. *Trichonema reflexum* (Eckl.) Steud., Nom. Bot. 2: 702 (1841), excl. syn. Thunb.

R. muirii N. E. Br. in Gdnrs' Chron. 92: 467 (1932). Type: Cape, Riversdale, *Muir* 4848 (K, holo.!).

Corm with hard tunics split into hard, bent basal teeth. *Leaves* filiform, up to 1 mm in diam. *Flowers* (20−) 25−38 mm long, magenta to pinkish lilac, drying to purple, sometimes white, with an orange-yellow cup, and often a violet-blue zone in the throat, outer segments on the backs often with 3−5 dark veins or paler lines and fine, feathered veining, or irregularly blotched. *Style* 7−14 mm long; stigmas not overtopping the anthers.

From Vanrhynsdorp to Riversdale.

Vouchers: *Barker* 7180 (NBG); *Taylor* 5106 (STE); *Leipoldt* 3193 (BOL); *Salter* 2678 (BOL).

This variety resembles *R. cruciata* var. *intermedia* (no. 60b) in its flowers, but the corms are distinct.

(c) var. **australis** *(Ewart) De Vos* in Jl S. Afr. Bot. Suppl. 9: 254 (1972). Type: Australia, near Melbourne, *Tovey* s.n. (MEL, lecto.!; BOL!).

R. cruciata var. *australis* Ewart in Proc. R. Soc. Vict. 19: 43 (Feb. 1907).

Trichonema cruciatum sensu Ker-Gawl. in Curtis's bot. Mag. 16: t. 575 (1802), excl. syn. Jacq.; Irid. Gen. 81 (1827), excl. syn. Jacq. and Thunb.; sensu Klatt in Linnaea 34: 662 (1865−66), excl. syn.

T. longifolium Salisb. in Trans. hort. Soc. Lond. 1: 316 (1812). *Bulbocodium longifolium* (Salisb.) Kuntze in Rev. Gen. 2: 701 (1891). *R. longifolia* (Salisb.) Bak. in Trans. Linn. Soc., Lond. 16: 89 (1877); in F. C. 6: 41 (1896). Iconotype: Curtis's bot. Mag. 16: t. 575 (1802).

R. minuta (L.) Eckl., Top. Verz. 19 (1827), partly, excl. syn. L. Type: Cape, C. B. S., *Ecklon* s.n. (S, holo.!).

R. cruciata var. *neglecta* Bég. in Bot. Jb. 38: 337 (March 1907); in Malpighia 23: 69 (1909). Syntypes: Cape, Vankamps (Camps) Bay, *Krauss* s.n. (G!; M!); without locality, *Brehm* s.n. (M!).

R. cruciata var. *parviflora* Bég. in Bot. Jb. 38: 337 (1907); in Malpighia 23: 69 (1909). Syntypes: Cape, C. B. S., *Zeyher* 4040 (G!; P!); near Vankamps (Camps Bay), *H. A. A. MacOwan* 1780 (B†; GRA!; SAM!; BM!; K!, P!); near Claremont, *Schlechter* 1567 (Z!).

R. rosea var. *parviflora* sensu G. J. Lewis in Adamson & Salter, Fl. Cape Penins. 223 (1950), non Bak. (1892 & 1896), nec Bég. (1909)? *R. rosea* var. *neglecta* sensu De Vos in Tydskr. Natuurwet. 5: 139 (1965), non Bég.

Corm as in var. *reflexa*. *Leaves* compressed cylindrical, 1−2,5 mm in diam., often spreading, with rather wide grooves. *Flowers* 15−25 mm long, pale lilac-pink or sometimes white, cup pale yellow, outer segments yellowish green or with 3−5 longitudinal lines on the backs. *Style* 7−10 mm long; stigmas not overtopping the anthers.

Widespread and very common from Calvinia to Port Elizabeth, often a weed along roadsides. Known as 'froetang' or 'knikkertjie'. Also introduced (?) into Tristan da Cunha, St Helena and Guernsey, as well as Australia where it has become a noxious weed known as onion grass or Guildford grass.

Vouchers: *Salter* 6824 (BOL); *Gillett* 1137 (BOL); *Barker* 1815 (NBG); *Acocks* 21481 (NBG; PRE; K).

Ewart's varietal epithet *australis* precedes those of Béguinot (*neglecta* and *parviflora*) by one month. This is not *R. rosea* var. *parviflora* Bak.

(d) var. **communis** *De Vos* in Jl S. Afr. Bot. Suppl. 9: 257 (1972). Type: Cape, Stellenbosch, near Papegaaisberg, *De Vos* 1099 (STE, holo.!).

Very closely related to var. *australis*, differing in the often more erect and slightly narrower leaves with slightly narrower grooves, and slightly larger flowers (22−) 25−35 mm long, often with a somewhat brighter colouring and a bluish zone in the throat.

From Clanwilliam to Humansdorp, sometimes a weed along roadsides.

Vouchers: *De Vos* 1093 (STE); *Bolus* 3746 (BOL; K); *Lewis* 1663 (SAM).

(e) var. **elegans** *(Klatt) Bég.* in Malpighia 23: 63 (1909); De Vos in Jl S. Afr. Bot. Suppl. 9: 258 (1972). Type: Cape, Doornhoogte, *Zeyher* 1602 (S, holo.!; GRA!; PRE!; SAM!; B!; G!; K, partly!).

R. elegans Klatt in Abh. Naturforsch. Ges. Halle 15: 400 (1882); in Dur. & Schinz, Consp. Fl. Afr. 5: 164 (1895), excl. cit. *Zeyher* 4043; Bak., Handb. Irid. 103 (1892); in F.C. 6: 42 (1896), partly, excl. var. and cit. *Zeyher* 4043; G. J. Lewis in Adamson & Salter, Fl. Cape Penins. 223 (1950), partly, excl. var.

Corm and *leaves* as in var. *reflexa*. *Flowers* 30−48 mm long, white with a golden yellow cup, outer perianth segments reddish purple or reddish green on the backs, irregularly marked or with a pale median line. *Style* 10-12 mm long; stigmas reaching halfway up the anthers to slightly above their tips.

Found on the Cape Peninsula and Cape Flats, and near Swellendam.

Vouchers: *Lewis* 59, 1599 (SAM); *Salter* 7637 (BOL; SAM); *Barker* 4116, 4794 (NBG); *De Vos* 1909 (STE).

Closely related to var. *rosea*, differing in flower colour and usually shorter styles.

60. **Romulea cruciata** *(Jacq.) Bak.* in J. Linn. Soc., Bot. 16: 89 (1877), partly, excl.

FIG. 17.−1, **Romulea cruciata** var. **cruciata**, habit, × 7/8; 1a, corm seen from the opposite side; 1b, bracts, outer (left) and inner (right); 1c, transverse section of leaf (*De Vos* 1620). 2, **R. eximia**, habit, × 3/4; 2a, mature capsules; 2b, outer (left) and inner (right) bract; 2c, outer perianth segment, lower face (*De Vos* 1687).

syn. *Trichonema cruciatum* Ker-Gawl.; *R. rosea* var. *speciosa* Bak. in F. C. 6: 42 (1896), partly as to syn. *Ixia cruciata;* Klatt in Abh. Naturforsch. Ges. Halle 15: 401 (1882); Bég. in Bot. Jb. 38: 335 (1907), partly, excl. var. 3, var. 4, syn. *R. rosea* and *Trichonema cruciatum* Ker-Gawl.; in Malpighia 23: 66 (1909), partly, excl. same vars & syns; De Vos in Jl S. Afr. Bot. Suppl. 9: 259, fig. 88 (1972), non Eckl., nec Lewis 1950. Iconotype: Jacq., Icon. Pl. Rar. 2: t. 290 (1790).

Plants 150—400 mm long. *Corm* obovoid, tunics with almost straight, acuminate basal teeth converging to a basal point. *Leaves* basal, filiform or X-shaped in transverse section, 140—400 × 0,8—4 mm, grooves narrow or wide. *Bracts* greenish or purplish red, inner with wide, brown, brown-streaked, or rarely colourless membranous margins. *Flowers* 22—35 (—42) mm long, magenta-pink to lilac-pink, with a dark blotch on each segment, cup yellow or orange-yellow, outer segments purplish pink or greenish on the backs or with 5 dark lines and fine, feathered veining. *Perianth tube* 3—5 mm long; segments 20—35 × 5—10 mm. *Filaments* 3—6 mm; anthers 4—8 mm, reaching less than halfway up the perianth. *Style* 9—11 mm; stigmas usually not overtopping the anthers. *Capsules* often shortly beaked, on erect or slightly spreading peduncles. *Chromosome no.* 2n=18.

Widely distributed in western, south-western and southern Cape districts from Vanrhynsdorp to Riversdale. Map 18.

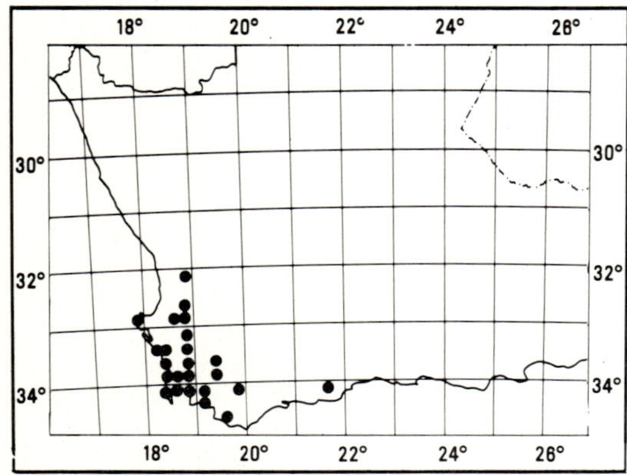

MAP 18.— **Romulea cruciata**

Related to *R. rosea* (no. 59) and distinguished by its corm with a pointed base and almost straight, basal teeth, its stamens not reaching halfway up the perianth, and by its slightly beaked capsules which have a somewhat delayed dehiscence and are borne on erect or slightly spreading peduncles. Two varieties occur which differ mainly in leaf width.

1 Widest leaves more than 1 mm in diameter, with wide grooves and sometimes with a narrow longitudinal ridge down the middle of each groove .... (a) var. *cruciata*
1 Widest leaves 1 mm or less in diameter, with narrow grooves .... (b) var. *intermedia*

(a) var. cruciata.

Ixia cruciata Jacq., Icon Pl. Rar. 2: t. 290 (1790); Coll. 5: 16 (1796); Willd., Sp. Pl. 1: 197 (1797); Poir., Encycl. Suppl. 3: 201 (1813); Roem. & Schult., Syst. Veg. 1: 377 (1817), excl. syn. Ker-Gawl. & Thunb. *Bulbocodium cruciatum* (Jacq.) Kuntze, Rev. Gen. 2: 700 (1891).

Trichonema cruciatum sensu Ait., Hort. Kew. 1: 83 (1810), partly, excl. syn. Ker-Gawl. & Mill.

Leaves (1—) 2—4 mm wide, often cruciform in transverse section, grooves wide, sometimes with a narrow longitudinal ridge in the middle of each wide groove. *Flowers* lilac-pink to magenta-pink, often with a blue zone in the throat. Capsules 10—14 mm long. Fig. 17: 1.

From Piketberg to the Cape Peninsula and Caledon.

Vouchers: *Schlechter* 5260 (BOL; GRA; Z); *Loubser* 975 (NBG); *Lewis* 1059 (SAM); *De Vos* 1620 (STE).

(b) var. intermedia *(Bég.) De Vos* in Jl S. Afr. Bot. Suppl. 9: 263, fig. 89 (1972). Type: Cape, *Zeyher* 4044 (G, holo.!; GRA!; K!; P!; S!; Z!).

R. intermedia Bég. in Bot. Jb. 38: 339 (1907); in Malpighia 23: 66 (1909).

R. ambigua Bég. in Bot. Jb. 38: 338 (1907), partly, excl. syn.; in Malpighia 23: 80 (1909), partly, excl. syn. and spec. *Ecklon* 703. Type: Cape, Clanwilliam, Olifants River near Brakfontein, *Schlechter* 10784 (B, lecto.!. GRA!; PRE!; G!; K!; S!).

Leaves filiform, 0,5—1 mm in diam., with narrow grooves. *Flowers* magenta-pink with violet blotches in the throat. *Capsules* 6—10 mm long.

From Vanrhynsdorp to the Cape Peninsula and to Riversdale.

Vouchers: *Lewis* 1347 (SAM); *Salter* 3557 (BOL); *Schlechter* 10784 (GRA; PRE; B; K; G; S); *De Vos* 1693 (STE).

61. Romulea eximia *De Vos* in Jl S. Afr. Bot. Suppl. 9: 267, figs 77 & 90 (1972). Type: Cape, Malmesbury, near Darling, *De Vos* 1687 (STE, holo.!).

R. speciosa sensu Bég. in Malpighia 23: 85 (1909), partly, excl. syn., non Bak.

Plants 250−450 mm long. *Corm* obovoid, with straight acuminate basal teeth converging to a basal point. *Leaves* basal, filiform, 250−450 × c. 1−1,5 mm, grooves narrow. *Bracts* purplish brown or greenish, inner with wide, brown membranous margins. *Flowers* 40−50 (−60) mm long, old-rose to dark old-rose, with a maroon blotch on each segment in the throat, cup pale yellow or greenish yellow, outer segments irregularly marked with red and greenish yellow on the backs. *Perianth tube* 5−8 mm long; segments 33−40 (−50) × 7−14 mm, the outer often slightly longer than the inner. *Filaments* 9−12 mm; anthers 7−12 mm long, often reaching less than halfway up the perianth. *Style* 18−20 mm; stigmas usually just below the anther tips. *Capsules* on suberect or slightly curved peduncles. *Chromosome no.* 2n=18. Fig. 17: 2.

From the south-western Cape district of Malmesbury (3318−AC, AD, BC, CB, DA).

Vouchers: *Compton* 1168/26 (BOL); *Bolus* 21262 (BOL); *Barker* 3839 (NBG); *Lewis* 1058 (SAM); *De Vos* 1687 (STE).

The large old-rose flowers with their maroon blotches resemble those of the sympatric *R. hirsuta* (no. 21) and *R. obscura* var. *blanda* (no. 56d), but *R. eximia* is distinguished by a pale greenish yellow perianth cup with dark blotches and corm with a pointed base and straight basal teeth.

6. Section **Spatalanthus**

Spatalanthus (*Sweet*) *Diels* in Natürl. PflFam. edn 2, 15a: 475 (1930), emend. De Vos. Type species: *R. monadelpha* (Sweet) Bak.

Spatalanthus Sweet, Brit. Fl.Gdn 3: t. 300 (1829), as genus; Bak. in F.C. 6: 37 (1896), as subgenus *Spathalanthus*.

Section *Bicarinatae* De Vos in Jl S. Afr. Bot. Suppl. 9: 274 (1972), nom. superfl.

Corm with tunics split into bent acuminate basal teeth curved over a rounded base. *Stem* short, hidden. *Foliage leaves* basal, glabrous, filiform or sometimes slightly swollen with widened rib margins, up to 5 mm in diam. *Bracts* large, firm, concave, green, with membranous margins, the outer with a stronger median vein, the inner two-keeled with two stronger veins. *Flowers* large, red, pink or yellow, variously marked with dark blotches in the throat. *Perianth tube* short, shallow, saucer-shaped. *Filaments* free or joined; anthers longer than the filaments, at first joined at the tips. *Stigmas* usually below the anther tips. *Capsules* on peduncles at first curved, later suberect.

From the western inland Cape districts of Calvinia and Sutherland at 1 300 — 1 700 m altitude.

Diels (1930) included in this section only *R. monadelpha* (no. 65) which is characterized by fused filaments. Four more taxa are now included which, notwithstanding free filaments, are closely related to *R. monadelpha*, corresponding to this species in their firm bracts, flowers with very shallow perianth tube, anthers longer than the filaments, and a chromosome number of 2n=26.

62. Romulea subfistulosa *De Vos* in Ann. Univ. Stell. 28A, 3: 66, fig. 3 (1952); in Jl S. Afr. Bot. Suppl. 9: 275, fig. 1 (1972). Type: Cape, near Sutherland, *Joubert & De Vos* 1585 (STE, holo.!).

Plants 120−250 mm long. *Corm* with curved basal teeth. *Stem* short or sometimes up to 90 mm long, hidden. *Leaves* somewhat swollen, arcuate or suberect, 100−250 × 1,5−5 mm, the rib margins often widened into 8 narrow longitudinal wings or ridges. *Peduncles* 50−150 mm long. *Bracts* green, outer with narrow membranous margins, inner with wider, minutely speckled, colourless membranous margins and small tips. *Flowers* 30−60 mm long, shiny carmine-rose, with a reddish black blotch on each segment, cup yellow with 6 dark lines, outer segments yellowish on the backs with 5−7 longitudinal lines and fine feathered veining. *Perianth tube* cup-shaped, 3−5,5 mm long; segments shortly oblanceolate,

25—50 × 10—17 mm. *Filaments* free, 4—6 mm, purple or yellow; anthers 8—11 mm long, bright yellow. *Style* 9—12 mm long.

Found in the Sutherland district and towards Middelpos in the Calvinia district on the Groot Roggeveld plateau (3320—AA, BC; 3221—AA).

Vouchers: *Marloth* 9658 (PRE; STE; B); *Hall* 200 (NBG; STE); *Oliver* 4413 (STE); *De Vos* 2177 (STE); *Joubert* in STE 30240.

Distinguished from *R. sabulosa* (no. 64) in its wide, almost fistulose leaves with wide stomatiferous grooves and in the colouring of the flowers. Herbarium specimens may be confused with *R. atrandra* (no. 48) but are distinguished by two-keeled inner bracts with two stronger veins and membranous margins narrowing suddenly to a very small scarious tip, and by peduncles which do not coil up in the fruiting stage.

63. **Romulea viridibracteata** *De Vos* in Jl S. Afr. Bot. Suppl. 9: 277, figs 83 & 92 (1972). Type: Cape, Clanwilliam, summit of Pakhuis Pass, *Salter* 3652 (BOL, holo.!; BM!; K!).

Plants 100—300 mm long. *Corm* with curved basal teeth. *Stem* short. *Leaves* filiform, 100—250 × 0,8—2 mm, curved or suberect, grooves narrow or wide. *Bracts* with narrow, dark brown, brown-dotted, or sometimes colourless membranous margins. *Flowers* 25—40 mm long, buttercup-yellow or ochre-yellow, with a purplish brown or black blotch on each segment, outer segments brown or purplish brown on the backs. *Perianth tube* shallow, saucer-shaped, 2—4 mm deep; segments 20—35 × 6—10 mm. *Filaments* free, 4—5 mm; anthers 6—9 mm long. *Style* 8—10 mm long.

From the Nieuwoudtville area in the Calvinia district and the Pakhuis mountains above Clanwilliam (3119—AC; 3219—AA).

Vouchers: *Leipoldt* 4423 (BOL), 483 (GRA); *De Vos* 2109, 1920 (STE); *Salter* 3652 (BOL; BM; K).

Related to *R. sabulosa* (no. 64) and differs mainly in its yellow, slightly smaller flowers with narrower perianth segments. The flower resembles those of *R. luteoflora* (no. 49) and *R. montana* (no. 11), but the species is readily distinguished by its firmer outer bracts and two-keeled inner bracts, and its wide, saucer-shaped perianth tube.

64. **Romulea sabulosa** *Schltr. ex Bég.* in Bot. Jb. 38: 334 (1907), excl. syn.; in Malpighia 23: 84 (1909), excl syn.; De Vos in Flower. Pl. Afr. 41: t. 1612 (1970); in Jl S. Afr. Bot. Suppl. 9: 279, figs 79 & 93 (1972). Syntypes: Cape, Olifants River near Clanwilliam, *Bergius* (B); Onder-Bokkeveld, Oorlogskloof, Calvinia, *Schlechter* 10964 (G, lecto.!; BOL!; GRA!; PRE!; B!; BM!; K!; S!; Z!).

Plants 120—400 mm long. *Corm* with curved basal teeth. *Leaves* filiform, 100—400 × c. 1 mm, grooves narrow. *Bracts* with narrow membranous margins which are usually colourless in the lower part and brown upwards. *Flowers* 30—50(—65) mm long, shiny scarlet or currant-red, each segment with a brownish black blotch on a greyish green, brownish or sometimes purplish blue background and below that a yellow V-shaped blotch, outer segments with 5—7 yellow veins and fine feathered veining on the backs. *Perianth tube* shallow, saucer-shaped, 2—4 mm deep, with a white, 6-pointed star-shaped blotch inside; segments subrhomboid-cuneate to obovate-cuneate, 25—40 (—55) × 12—20 (—25) mm. *Filaments* free, 3—5 mm, cream or dark; anthers 8—12 mm long, purplish or yellow. *Style* 7—12 mm; stigmas halfway up the anthers. Fig. 18:2.

Very common in the Nieuwoudtville area in the Calvinia district, on the Bokkeveld Mountains escarpment (3119—AC, CA).

Vouchers: *Acocks* 19022 (PRE; STE); *De Vos* 2024 (STE); *Lewis* 5831 (NBG); *Marloth* 5566; *Leipoldt* 3822 (BOL).

Known as 'satynblom' or 'syblom'; distinguished by its shiny, almost bell-shaped red flowers with characteristic markings, rigid green bracts, of which the inner are two-keeled, and its free filaments.

65. **Romulea monadelpha** *(Sweet)* *Bak.*, Handb. Irid. 104 (1892); in F.C. 6: 43 (1896); Klatt in Dur. & Schinz, Consp. Fl. Afr. 5: 165 (1895); Bég. in Malpighia 23: 83 (1909), pro syn.; De Vos in Jl S. Afr. Bot. 36: 1 (1970), ibid. Suppl. 9: 283, figs 81 & 94 (1972); Veld & Flora 1: 37 (1971). Iconotype: Sweet, Brit. Fl. Gdn 3: t. 300 (1829).

FIG. 18.—1, **Romulea monadelpha**, habit, × 7/8; 1a, outer perianth segment, lower face; 1b, inner segment, upper face; 1c, perianth base, stamens and pistil; 1d, outer (left) and inner (right) bract; 1e, almost mature capsule; 1f, transverse section of leaf (*De Vos* 1991). 2, **R. sabulosa**, perianth base, stamens and pistil; 2a, inner perianth segment, upper face.

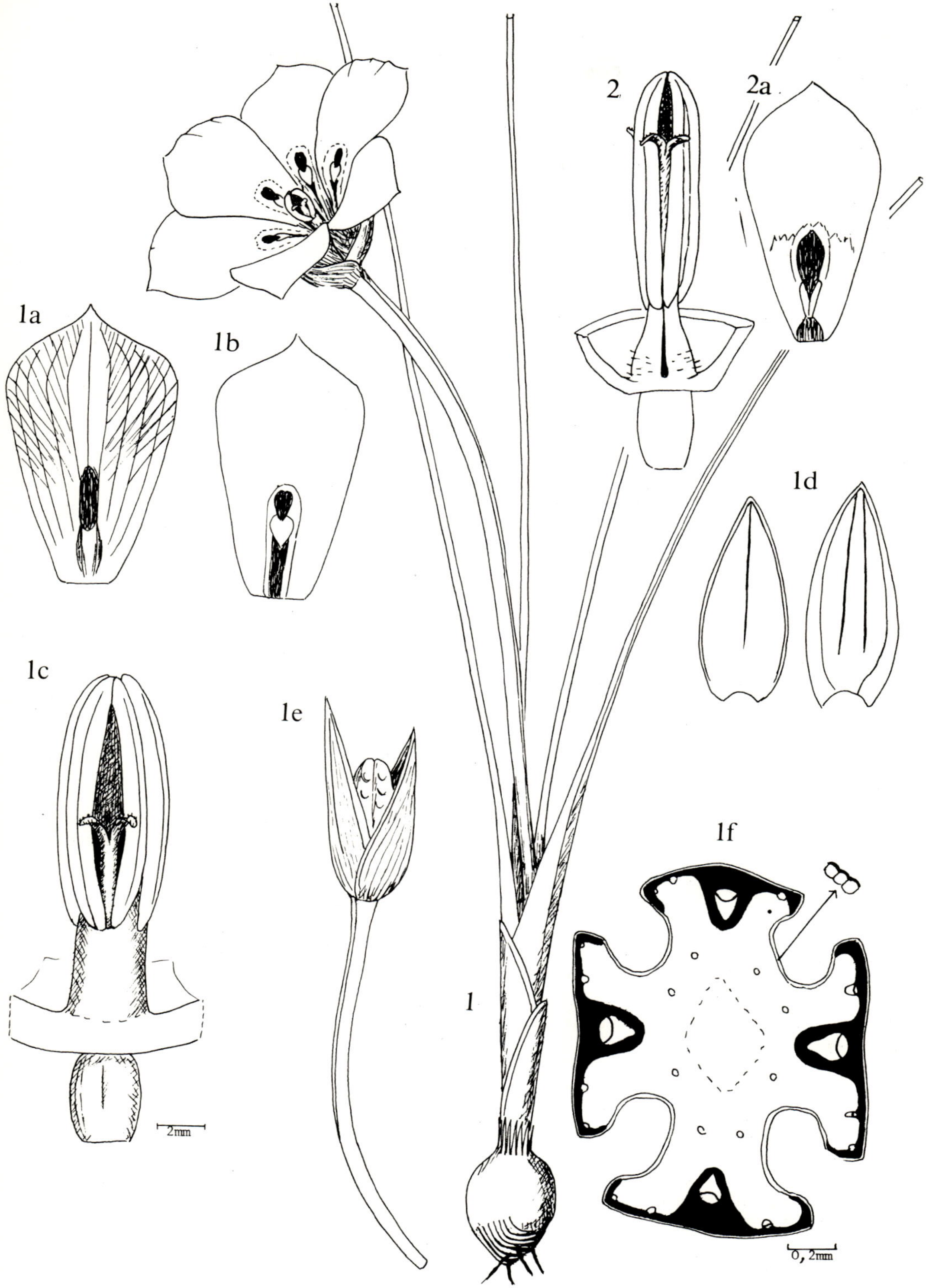

1a
1b
1c
1d
1e
1f
1
2
2a
2mm
0,2mm

Trichonema monadelphum Sweet, Hort. Brit. edn 2, 399 (1830). *Spatalanthus speciosus* Sweet, Brit. Fl. Gdn 3: t. 300 (1829), ibid (1837); Hort. Brit. 668 (1839); Bak. in J. Linn. Soc., Bot. 16: 104 (1877); Klatt in Abh. Naturforsch. Ges. Halle 15: 386 (1882). Type as for *R. monadelpha.*

Very closely related to *R. sabulosa* (no. 64) from which it differs as follows: *Flowers* 30−45(−55) mm long, deep claret-red, each segment with a black blotch on a blue or purplish grey or sometimes pale yellow background. *Filaments* joined, forming a short, stout, shiny-black column; anthers 10−15 mm long. *Style* 8−12 mm long. Fig. 18:1.

In the Nieuwoudtville area in the Calvinia district, on the Bokkeveld Mountain escarpment (3119−AC, BC).

Vouchers: *Lewis* 2160 (STE); *De Vos* 1926 (STE); *Hardy* 64; *Middlemost* 2160 (NBG); *Goldblatt* 260 (BOL).

This rare species is also known as 'satynblom' or 'syblom' on account of its shiny perianth. It is interfertile with *R. sabulosa* (no. 64), the F_1 hybrids resembling the latter in their free filaments (De Vos, l.c. 1970). Probably some specimens, distinguished as *R. sabulosa*, are hybrids.

66. **Romulea vanzyliae** *De Vos* in Jl S. Afr. Bot. Suppl. 9: 284 (1972). Type: Cape, without locality, cult. *Van Zyl* in BOL 24349 (BOL, holo.!; PRE!).

Very closely related to *R. sabulosa* (no. 64) and *R. subfistulosa* (no. 62) and previously treated as a putative hybrid; differing from the former as follows: *Leaves* somewhat swollen, with rib margins often widened, up to 5 mm wide. *Flowers* pinkish red, each segment with a dark blotch on a pale violet zone, and below that a yellow blotch. *Anthers* up to 14 mm long.

From the Calvinia district towards Nieuwoudtville (3119−AC), in a damp locality.

Vouchers: *Schmidt* 341; *De Vos* 2378 (STE); BOL 24349 (BOL; PRE).

More work is necessary on this rare taxon.

2. Subgenus **Lomurea**

Lomurea *De Vos* in Jl S. Afr. Bot. Suppl. 9: 285 (1972). Type species: *R. syringodeoflora* De Vos.

Flowers salver-shaped, magenta, rose or pale violet. *Perianth tube* largely tubular, 11 −70 mm long, longer than the segments; segments spreading horizontally, generally narrowly elliptical, obtuse, up to 17 mm long. *Stamens* inserted in the upper part of the perianth tube; filaments glabrous. *Capsules* produced just above ground-level on suberect peduncles.

The subgenus comprises three not very closely related species from the mountainous regions of the western Cape Province, namely the Bokkeveld, Cedarberg and Roggeveld mountain ranges at altitudes of c. 1 500 m.

7. Section **Lomurea**

Lomurea *De Vos* in Jl S. Afr. Bot. Suppl. 9: 287 (1972). Type species: *R. syringodeoflora* De Vos.

Corm with a rounded or pointed base, with tunics split at base into acuminate, bent or almost straight, ungrooved teeth. *Leaves* several, compressed cylindrical, arcuate. *Bracts* largely green. *Perianth tube* 15−50 mm long, narrowly tubular, widened at top; segments widely patent. *Style* 22−65 mm long. *Capsules* shortly cylindrical.

67. **Romulea syringodeoflora** *De Vos* in Ann. Univ. Stell. 28A, 3: 74, fig. 6 (1952); in Jl S. Afr. Bot. Suppl. 9: 287, fig. 95 (1972). Type: Cape, flats near Sutherland, *De Vos* 1587 (STE, holo.!).

Plants 120−200 mm long. *Corm* subglobose or ovoid, with acuminate, usually bent, basal teeth. *Stem* short, hidden. *Leaves* compressed cylindrical, arcuate or recurved, 120−200 × 1−2,5 mm, grooves wide, rib margins ciliate or glabrescent. *Bracts* green with colourless, membranous margins and tips, reaching halfway or higher up the perianth tube. *Flowers* 30—40 mm long, magenta-pink, with a maroon V-shaped mark on each segment near the base, outer segments striped maroon and yellow on the back. *Perianth tube* 15—22 mm long, tubular, widened slightly at the top; segments narrowly elliptical or oblanceolate, spreading horizontally, obtuse, 10—17 × 3—5 mm. *Filaments* 4—5 mm, pale; anthers 4—6 mm long, purple, pollen yellow, purple or rust-coloured. *Style* 22—30 mm long; stigmas reaching below to just above the anther tips. *Chromosome no.* 2n=20.

Found only on the Roggeveld plateau near Sutherland and towards its edge (3220—BC, —DA).

Vouchers: *Marloth* 9644 (PRE; STE; B); *Acocks* 17797 (PRE; K); *Joubert in STE* 27156, 27159; *De Vos* 2059 (STE).

Readily distinguished by its salver-shaped, long-tubed flowers with spreading obtuse segments, and corm with a rounded base.

68. **Romulea hantamensis** (*Diels*) Goldbl. in Flower. Pl. Afr. 41:t. 1613 (1970); De Vos in Jl S. Afr. Bot. Suppl. 9: 289, fig. 96 (1972).

Lapeirousia hantamensis Diels in Bot. Jb. 44: 116

(1910). Type: Cape, Calvinia, westlich der Hantams-Berge, Gipfelfläche, *Diels* 732 (B, holo.!).

Plants 70—150 mm long. *Corm* with straight acuminate teeth converging to a pointed base. *Leaves* compressed cylindrical, arcuate or recurved, glabrous, 7—15 × 1—1,5 mm, grooves rather narrow. *Bracts* largely green, purple towards the base, ensheathing the lower part of the perianth tube, inner with wide, colourless or brown-speckled membranous margins. *Flowers* 60—75 mm long, bright magenta, with a purplish black blotch above the middle of each segment, and below that a long white blotch with three dark lines, outer segments purple striped on the backs. *Perianth tube* 35—70 mm long, tubular, widened slightly at the top; segments elliptical, spreading horizontally, 10—14 × 3—5 mm. *Filaments* 3 mm, purple; anthers 3—5 mm, purple striped; pollen yellow. *Style* 60—65 mm; stigmas at or just above the anther tips. *Chromosome no.* 2n=30. Fig. 19:1.

Found only on the Hantam Mountain range above Calvinia at c. 1 500 m altitude (3119—BC, —BD).

Vouchers: *Goldblatt* 276 (BOL; STE), 429; *Diels* 732 (B); *Thompson* 2318 (STE).

Resembles *R. syringodeoflora* (no. 67) somewhat but differs in its corm, its much longer perianth tube, longer style, differently marked perianth segments, and in chromosome number.

8. Section **Stellanthe**

Stellanthe *De Vos* in Jl. S. Afr. Bot. Suppl. 9: 291 (1972). Type species: *R. stellata* De Vos.

Corm with an oblique crescent-shaped basal ridge. *Leaves* 1(—2), filiform. *Bracts* submembranous. *Perianth tube* 11—17 mm long, narrowly tubular, longer than the segments; segments widely patent. *Style* 15—20 mm long. *Capsules* ellipsoid.

69. **Romulea stellata** De Vos in Jl S. Afr. Bot. Suppl. 9: 291, fig. 97 (1972). Type: Cape, Clanwilliam, Pakhuis Pass beyond the summit, *De Vos* 2171 (STE, holo.!).

Plants 100—130 mm long. *Corm* with minute parallel fibrils on a crescent-shaped basal ridge. *Stem* short, hidden. *Leaves*

1—2, filiform, flexuose or suberect, 100—130 × 0,5—0,8 mm, grooves very narrow, the adaxial groove open almost to the leaf tip. *Bracts* submembranous, reddish brown or greenish in the upper part, inner with narrow membranous margins. *Flowers* 20—30 mm long, pale purple or pale violet, darker violet in the throat, outer segments mottled purple on the back; *Perianth tube*

FIG. 19.—1, **Romulea hantamensis**, habit, × 1; 1a, outer perianth segment, upper and lower face; 1b, outer (left) and inner (right) bract; 1c, perianth tube, stamens and pistil; 1d, transverse section of leaf (*Diels* 732, *Goldblatt* 276). 2, **R. stellata**, habit, × 1; 2a, outer perianth segment: upper (left) and lower (right) face; 2b, outer (left) and inner (right) bract; 2c, ripening capsule; 2d, transverse section of leaf (*De Vos* 2171).

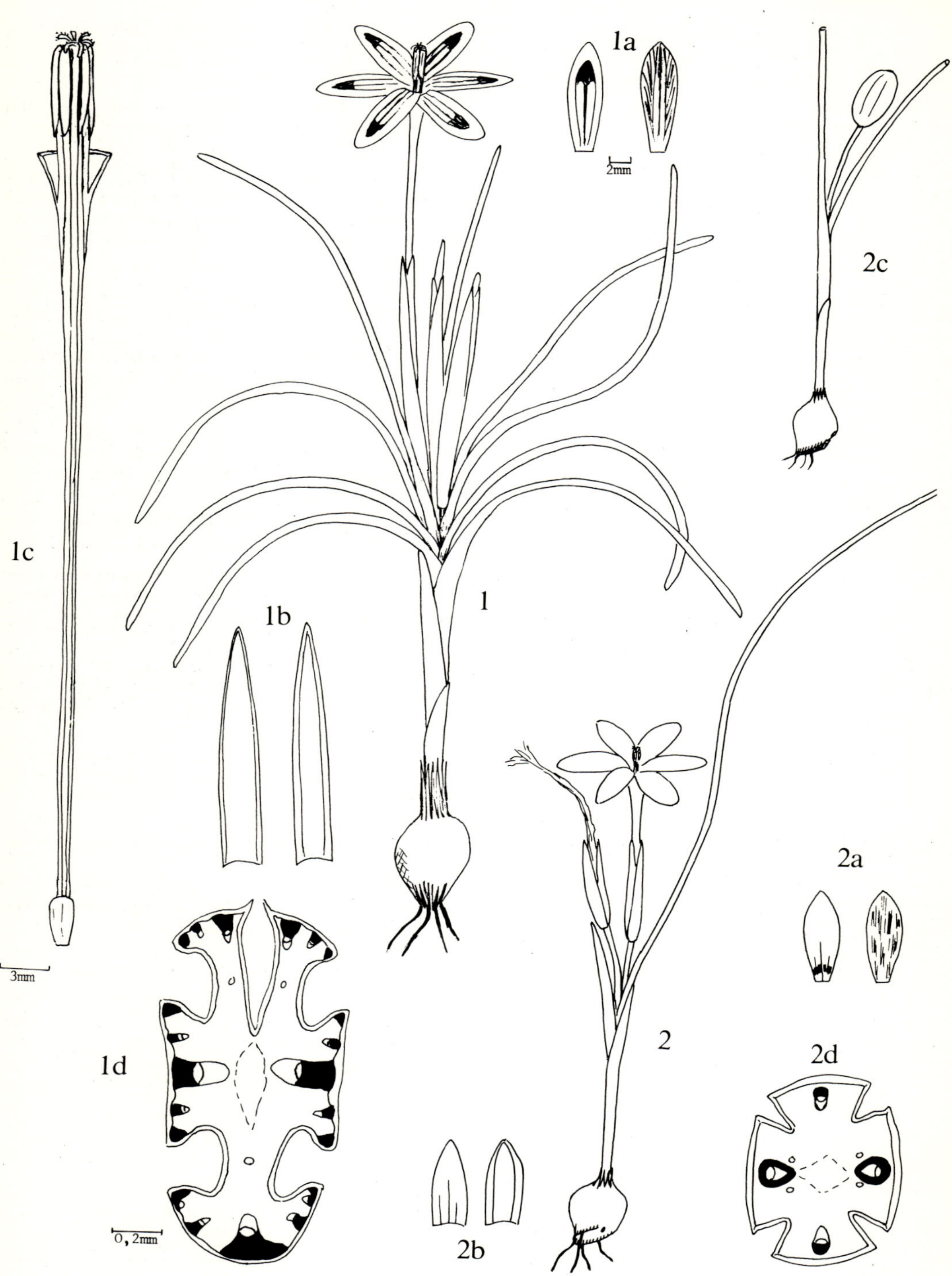
1a
2mm
1c
1b
1d
3mm
0,2mm
1
2b
2
2c
2a
2d

11−17 mm long, narrowly tubular; segments elliptical, 7−11 × 2− 4mm. *Filaments* 2−3,5 mm; anthers 2−3 mm long, violet. *Style* 15 − 20 mm; stigmas at the anther tips. Fig. 19: 2.

Only known from the Gifberge near Vanrhynsdorp and the Pakhuis area in the Cedarberg Mountains of the Clanwilliam district; in moist depressions on rock layers (3118−DC; 3219−AA, −AC).

Vouchers: *Salter* 2452, 2460 (BOL); *Oliver* 4935 (STE), s.n., 30-5-65 (NBG); *De Vos* 2171 (STE).

This small, early flowering (May to July), montane species is apparently not closely related to other species of *Romulea*. It is distinguished by its small salver-shaped, star-like flowers with a long perianth tube, by one or two very slender filiform leaves, and by its corm with a crescent-shaped ridge.

Species insufficiently known

Ixia bulbocodioides F. Delaroche, Descr. Pl. 19 (1766). Type not found. The description of the corm fits *Romulea triflora* and the leaf description *R. flava*.

I. fugax Salisb., Prodr. 34 (1796). Type not found. Judging by the description it is a species of *Romulea*.

I. fugax Hornem., Hort. Hafn. 50 (1813). The type at C is too poor for identification and the description is inadequate.

Romulea barbata Bak. in J. Bot. 5: 236 (1876). The holotype, a Herbert manuscript figure sub *Trichonema barbatum* W. Herb. in the Lindley Library, R. Hort. Soc., London, is not a variety of *R. rosea* as Baker (1892, 1896) stated. It is probably a horticultural form.

R. dielsii Bég. in Malpighia 23: 96 (1909): *R. hirsuta* × *R. cruciata*. Type: *Diels* 146 in B not found. Hybridization between the cited species is unlikely.

Trichonema pudicum Ker-Gawl. in Curtis's bot. Mag. 31: t. 1244 (1810). *Ixia pudica* (Ker-Gawl.) Roem. & Schult., Syst. Veg. 1: 375 (1817). *R. pudica* (Ker-Gawl.) Bak. in J. Linn. Soc., Bot. 16: 89 (1877). *R. rosea* var. *pudica* (Ker-Gawl.) Bak. in F.C. 6: 42 (1896), partly. Iconotype: Curtis's bot. Mag. 31: t. 1244

(1810). No species of *Romulea* has been found which fits this figure, the nearest being *R. amoena* or *R. hirsuta* var. *framesii*. The figure does not fit Solander's specimen of *Ixia pudica* in BM which is *R. flava*.

R. speciosa (Ker-Gawl.) Bak. in J. Linn. Soc. Bot. 16: 89 (1877). *Trichonema speciosum* Ker-Gawl. in König & Sims, Ann. Bot. 1: 223 (1805). *R. rosea* var. *speciosa* (Ker-Gawl.) Bak., Handb. Irid. 103 (1892). Iconotype: *Ixia bulbocodium* var. *flore speciosissima* Andr., Bot. Rep. 3: 170 (1801). This figure is nearest to *R. dichotoma* (Thunb.) Bak. *T. speciosum* Ker-Gawl. in Curtis's bot. Mag. 36: t. 1476 (1812) is *R. neglecta* (Schultes) De Vos.

R. tubata Klatt in Abh. Naturforsch. Ges. Halle 15: 401 (1882). The holotype, *Drège* 2636 in Herb. Lübeck, was not found. Specimens labelled *Drège* 2636 in other herbaria (G, K, OXF, P and S) have flowers with a short perianth tube and do not fit Klatt's description. They are not considered to be isotypes. From the two sketches in S it is impossible to determine the species (or even the genus).

Excluded species

Romulea spiralis (Burch.) Bak. in J. Linn. Soc., Bot. 16: 90 (1877); in F.C. 6: 40 (1896). Type: Cape, near Sutherland, *Burchell* 1356 (K, holo.!). This is **Geissorhiza spiralis** (Burch.) De Vos in Jl S. Afr. Bot. Suppl. 9: 295 (1972), related to *G. corrugata* Klatt.

R. zeyheri Eckl., Top. Verz. 19 (1827), nom. nud. The Ecklon specimen in S is a **Geissorhiza** species and is not *R. zeyheri* (Bak.) Bég. which is **R. hirsuta** var. **zeyheri.**

Trichonema humile (Thunb.) Ker-Gawl. in Curtis's bot. Mag. 16: sub t. 575 (1802) is **Geissorhiza humilis** (Thunb.) Ker-Gawl. in König & Sims, Ann. Bot. 1: 224 (1805).

T. longitubum Klatt in Linnaea 34: 665 (1865-66) is **Syringodea longituba** (Klatt) Kuntze.

T. ornithogaloides (Licht. ex Roem. & Schult.) A. Dietr., Syn. Pl. 1: 159 (1839) is **Geissorhiza ornithogaloides** Klatt.

T. quadrangulum Sweet, Hort. Brit. 399 (1827) is **Gladiolus quadrangulus** (Delaroche) Barnard.

INDEX